Lecture Notes in Artificial Intelligence   7477

Subseries of Lecture Notes in Computer Science

LNAI Series Editors

Randy Goebel
University of Alberta, Edmonton, Canada
Yuzuru Tanaka
Hokkaido University, Sapporo, Japan
Wolfgang Wahlster
DFKI and Saarland University, Saarbrücken, Germany

LNAI Founding Series Editor

Joerg Siekmann
DFKI and Saarland University, Saarbrücken, Germany

Nadia Mana
Friedhelm Schwenker
Edmondo Trentin (Eds.)

Artificial Neural Networks in Pattern Recognition

5th INNS IAPR TC 3 GIRPR Workshop, ANNPR 2012
Trento, Italy, September 17-19, 2012
Proceedings

 Springer

Volume Editors

Nadia Mana
Fondazione Bruno Kessler (FBK)
38123 Trento, Italy
E-mail: mana@fbk.eu

Friedhelm Schwenker
University of Ulm
Institute of Neural Information Processing
89069 Ulm, Germany
E-mail: friedhelm.schwenker@uni-ulm.de

Edmondo Trentin
Università di Siena
Dipartimento di Ingegneria dell'Informazione
53100 Siena, Italy
E-mail: trentin@dii.unisi.it

ISSN 0302-9743   e-ISSN 1611-3349
ISBN 978-3-642-33211-1   e-ISBN 978-3-642-33212-8
DOI 10.1007/978-3-642-33212-8
Springer Heidelberg Dordrecht London New York

Library of Congress Control Number: 2012945738

CR Subject Classification (1998): I.2.6, I.5.1-3, I.4, J.3, H.5

LNCS Sublibrary: SL 7 – Artificial Intelligence

Preface

This 5th INNS IAPR TC3 GIRPR Workshop on Artificial Neural Networks for Pattern Recognition (ANNPR 2012), whose proceedings are presented in this volume, endeavored to bring together recent novel research in this area and to provide a forum for further discussion. The workshop was held at the Fondazione Bruno Kessler (FBK) in Trento, Italy, during September 17–19, 2012.

ANNPR 2012 was supported by the International Neural Network Society (INNS), by the International Association for Pattern Recognition (IAPR), by the IAPR Technical Committee on Neural Networks and Computational Intelligence (TC3), and by the Gruppo Italiano Ricercatori in Pattern Recognition (GIRPR). IAPR-TC3 is one of the 20 Technical Committees of IAPR, focusing on the application of computational intelligence to pattern recognition.

The workshop featured regular oral presentations and a poster session, plus three precious IAPR invited speeches, namely:

- "Developmental Vision Agents" given by Marco Gori of the Università di Siena, Dip. di Ingegneria dell'Informazione
- "NeuCube EvoSpike Architecture for Spatio-Temporal Modelling and Pattern Recognition of Brain Signals" given by Nikola (Nik) Kasabov of the AUT - Auckland University of Technology
- "Classifier Fusion with Belief Functions" given by Günther Palm of the University of Ulm, Institute of Neural Information Processing

It is our firm conviction that all the papers, included in the present book, are of high quality and significance to the areas of neural network-based and machine learning-based pattern recognition. We sincerely hope that readers of this volume may, in turn, enjoy it and get inspired from its different contributions.

We would like to acknowledge the fact that the organization of the workshop moved its first steps within the framework of the Vigoni Project for international exchanges between the universities of Siena (Italy) and Ulm (Germany). Also, we wish to acknowledge the generosity of the ANNPR 2012 sponsors: INNS, IAPR, IAPR-TC3, GIRPR, the University of Ulm, the Dipartimento di Ingegneria dell'Informazione (DII) of the University of Siena, and the Fondazione Bruno Kessler which hosted the event.

We are grateful to all the authors who submitted a paper to the workshop. Special thanks to the local chairs and organization staff, in particular to Oswald Lanz, Stefano Messelodi and Moira Osti. The contribution from the members of the Program Committee in promoting the event and reviewing the papers is gratefully acknowledged. Finally, we wish to express our gratitude toward Springer for publishing these proceedings within their LNCS/LNAI series, and for their constant support.

July 2012

Nadia Mana
Friedhelm Schwenker
Edmondo Trentin

Organization

Organizing Committee

Nadia Mana	Fondazione Bruno Kessler, Trento, Italy
Friedhelm Schwenker	University of Ulm, Germany
Edmondo Trentin	University of Siena, Italy

Program Committee

Shigeo Abe (Japan)
Amir Atiya (Egypt)
Erwin Bakker (The Netherlands)
Yoshua Bengio (Canada)
Ludovic Denoyer (France)
Neamat El Gayar (Canada)
Antonino Freno (France)
Markus Hagenbuchner (Australia)
Barbara Hammer (Germany)
Tom Heskes (The Netherlands)
Lakhmi Jain (Australia)
Nik Kasabov (New Zealand)
Hans A. Kestler (Germany)

Oswald Lanz (Italy)
Marco Loog (The Netherlands)
Simone Marinai (Italy)
Stefano Messelodi (Italy)
Heiko Neumann (Germany)
Erkki Oja (Finland)
Günther Palm (Germany)
Lionel Prevost (France)
Raul Rojas (Germany)
Stefan Scherer (USA)
Alessandro Sperduti (Italy)
Ah-Chung Tsoi (Macau)
Ian Witten (New Zealand)

Local Arrangements

Oswald Lanz
Stefano Messelodi
Moira Osti

Sponsoring Institutions

International Neural Network Society (INNS)
International Association for Pattern Recognition (IAPR)
Technical Committee 3 (TC3) of the IAPR
Gruppo Italiano Ricercatori in Pattern Recognition (GIRPR)
Fondazione Bruno Kessler (FBK), Trento, Italy
University of Ulm, Germany
DII, University of Siena, Italy

Table of Contents

Learning Algorithms

Applications

Invited Paper

How to Quantitatively Compare
Data Dissimilarities
for Unsupervised Machine Learning?

Bassam Mokbel[1], Sebastian Gross[2], Markus Lux[1],
Niels Pinkwart[2], and Barbara Hammer[1]

[1] CITEC Centre of Excellence, Bielefeld University, Germany
[2] Computer Science Institute, Clausthal University of Technology, Germany
bmokbel@techfak.uni-bielefeld.de

Abstract. For complex data sets, the pairwise similarity or dissimilarity of data often serves as the interface of the application scenario to the machine learning tool. Hence, the final result of training is severely influenced by the choice of the dissimilarity measure. While dissimilarity measures for supervised settings can eventually be compared by the classification error, the situation is less clear in unsupervised domains where a clear objective is lacking. The question occurs, how to compare dissimilarity measures and their influence on the final result in such cases. In this contribution, we propose to use a recent quantitative measure introduced in the context of unsupervised dimensionality reduction, to compare whether and on which scale dissimilarities coincide for an unsupervised learning task. Essentially, the measure evaluates in how far neighborhood relations are preserved if evaluated based on rankings, this way achieving a robustness of the measure against scaling of data. Apart from a global comparison, local versions allow to highlight regions of the data where two dissimilarity measures induce the same results.

1 Introduction

In many application areas, data are becoming more and more complex such that a representation of data as finite-dimensional vectors and their treatment in terms of the Euclidean distance or norm is no longer appropriate. Examples include structured data such as bioinformatics sequences, graphs, or tree structures as they occur in linguistics, time series data, functional data arising in mass spectrometry, relational data stored in relational databases, etc. In consequence, a variety of techniques has been developed to extend powerful statistical machine learning tools towards non-vectorial data such as kernel methods using structure kernels, recursive and graph networks, functional methods, relational approaches, and similar [9,12,5,27,6,26,10,11]. One very prominent way to extend statistical machine learning tools is offered by the choice of problem-specific measures of data proximity, which can often directly be used in machine learning tools based on similarities, dissimilarities, distances, or kernels. The latter include popular techniques such as the support vector machine, other

N. Mana, F. Schwenker, and E. Trentin (Eds.): ANNPR 2012, LNAI 7477, pp. 1–13, 2012.

kernel approaches such as kernel self-organizing maps or kernel linear discriminant analysis, or distance-based approaches such as k-nearest neighbor techniques or distance-based clustering or visualization, see e.g. [23]. Here, we are interested in dissimilarity-based approaches in general, treating metric distances as a special case of (non-metric) dissimilarities.

With the emergence of more and more complex data structures, several dedicated structure metrics have become popular. Classical examples include alignment for sequences in bioinformatics [22], shape distances [21], or measures motivated by information theory [4]. Often, there exists more than one generic possibility to encode and compare the given data. In addition, dissimilarity measures often come with parameters, the choice of which is not clear a priori. Hence, the question occurs how to choose an appropriate metric in a given setting. More generally, how can we decide whether a change of the metric or its parameters changes the data representation which is relevant for the subsequent machine learning task? Are there possibilities to compare whether and, if so, in which regions two metrics differ if used for machine learning?

Many approaches which are used in machine learning for structures have been proposed in the supervised domain. Here, a clear objective of the task is given by the classification or regression error. Therefore, it is possible to evaluate the difference of dissimilarities by comparing the classification error obtained when using these different data representations. A few extensive comparisons how different dissimilarities influence the outcome have been conducted; see, e.g. [18] for the performance of different dissimilarities for content-based image retrieval, [19] for an according study in the symbolic domain, [2] for the comparison of distances for probability measures, or [3] for the performance of classifiers on differently preprocessed dissimilarities to arrive at a valid kernel.

The situation is less clear when dealing with unsupervised domains. Unsupervised learning is essentially ill-posed and the final objective depends on expert evaluation. The primary mathematical goal is often to cluster or visualize data, such that an underlying structure becomes apparent. Quite a few approaches for unsupervised learning for structures based on general dissimilarities have been proposed in the past: kernel clustering techniques such as kernel self-organizing maps (SOM) or kernel neural gas (NG) [34,24] or relational clustering such as proposed for fuzzy-k-means, SOM, NG, or the generative topographic mapping (GTM) [13,7,8]. Further, many state-of-the art nonlinear visualization techniques such as t-distributed stochastic neighbor embedding are based on pairwise dissimilarities rather than vectors [31,15].

In this contribution, we will investigate how to compare dissimilarity measures with regard to their influence on unsupervised machine learning tasks, and discuss different possibilities in Sec. 2. Thereafter, we will focus on a principled approach independent of the chosen machine learning technique, rather we will propose a framework which compares two dissimilarity measures based on their induced neighborhood structure in Sec. 3. This way, it is possible to decide prior to learning whether and, if so, in which regions two different dissimilarity measures or different choices of parameters lead to different results, which we will demonstrate on examples in Sec. 4 and 5, concluding with a discussion in Sec. 6.

2   How to Compare Dissimilarity Measures?

We assume that data $\mathbf{x}_i$ are sampled from some underlying data space. These data are input to an unsupervised machine learning algorithm by means of pairwise comparisons $d_{ij} = d(\mathbf{x}_i, \mathbf{x}_j)$. These values constitute dissimilarities, as given by the squared Euclidean distance, provided data are vectorial. We assume that d refers to a general dissimilarity measure for which Euclidean properties are not necessarily guaranteed, maybe even the constraints of a metric are violated. Note, that the dual situation, similarities or kernels, can easily be transferred to this setting, see [23].

Interestingly, albeit the chosen dissimilarity structure crucially determines the output of any machine learning algorithm based thereon, no framework of how to compare different dissimilarities for unsupervised domains is commonly accepted in the literature. The question occurs what is the relevant information contained in a dissimilarity which guides the output of such an algorithm? Interestingly, even slight changes of the dissimilarity such as a shift can severely influence the result of an unsupervised algorithm, as shown in [8]. Apart from generic mathematical considerations, indications for the answer to this question may be taken from attempts to formalize axioms for unsupervised learning [1,17,33,14]. Here, guidelines such as scale-invariance, rank-invariance, or information retrieval perspectives are formalized. Now, we formalize and discuss different possibilities how to compare dissimilarity measures. We assume that pairwise dissimilarities d_{ij}^1 and d_{ij}^2, which are to be compared, are given.

Matrix Comparison: The pairwise dissimilarities d_{ij}^1 and d_{ij}^2 give rise to two square matrices D_1 and D_2 respectively, which could directly be compared using some matrix norm. This possibility, however, is immediately ruled out when considering standard axioms for clustering [1], for example. One natural assumption is scale-invariance of the unsupervised learning algorithm. Scaling the matrix, however, does affect the resulting matrix norm. More generally, virtually any matrix norm severely depends on specific numeric choices of the representation rather than the global properties of the data.

Induced Topology: An alternative measure which ignores numerical details but focuses on basic structures could be connected to the mathematical set-theoretic topology of a data space. Every distance measure induces a topology. Hence, it is possible to compare whether the topological structure induced by two metrics is equivalent. In mathematics, two metrics are called topologically equivalent if the inequality $c \cdot d^1(\mathbf{x}_i, \mathbf{x}_j) \leq d^2(\mathbf{x}_i, \mathbf{x}_j) \leq c' \cdot d^1(\mathbf{x}_i, \mathbf{x}_j)$ holds for all $\mathbf{x}_i, \mathbf{x}_j$ for some constants $0 < c \leq c'$, since they induce the same topology in this case. It can easily be shown that any two metrics in a finite-dimensional real vector space are topologically equivalent. However, this observation shows that this notion is not appropriate to compare metrics with respect to their use for unsupervised learning: topologically equivalent metrics such as the standard Euclidean metric and the maximum-norm yield qualitatively different clusters in practical applications, as we will demonstrate in an example in Sec. 4.

Rank Preservation: One axiom of clustering, as formalized in [1], is the invariance to rank-preserving distortions. Indeed, many clustering or visualization techniques take into account the ranks induced by the given dissimilarity measure only, this way achieving a high robustness of the results. Examples include algorithms based on winner-takes-all schemes or extensions such as vector quantization, NG, SOM, or similar approaches. Also, many visualization techniques try to preserve local neighborhoods as measured by the rank of data. How can rank-preservation be evaluated quantitatively? One way is to transform the matrices D_1 and D_2 into rank matrices, i.e. matrices which contain permutations of the numbers $\{0, \ldots, N-1\}$, N being the number of data points. Then, these two matrices could be compared by their column-wise correlation. However, usually the preservation of all ranks is not as critical as the preservation of a local neighborhood for most machine learning tools, such that different scales of the neighborhood size should be taken into account. In Sec. 3, we will explain the co-ranking framework which can be seen as a way to observe this rank-preservation property according to various neighborhood sizes of interest.

Information Retrieval Based Comparison: Information retrieval constitutes a typical application area for unsupervised learning. Therefore a comparison of dissimilarity measures based on this perspective would be interesting. Assume a user queries a database for the neighborhood of $\mathbf{x}_i$. What is the precision/recall, if d^2 is used instead of d^1? When defining the notion of neighborhood as the K nearest neighbors, precision and recall for a query $\mathbf{x}_i$ are both given by the term $|\{\mathbf{x}_j \mid d^1(\mathbf{x}_i, \mathbf{x}_j) \leq K \wedge d^2(\mathbf{x}_i, \mathbf{x}_j) \leq K\}|$ normalized by K. Summing over all $\mathbf{x}_i$ and dividing by N yields an average of all possible queries. In fact, this instantiation of a quality measure coincides with an evaluation within the co-ranking framework which will be introduced in Sec. 3.

3   The Co-ranking Framework

One very prominent tool in unsupervised learning is given by nonlinear dimensionality reduction and visualization [15]. Although many of the most relevant nonlinear dimensionality reduction methods have been proposed in the last years only, the question of what are appropriate quantitative evaluation tools is still widely unanswered. Interestingly, as reported in [32], a high percentage of publications on data visualization evaluates results in terms of visual impression only – about 40% out of 69 papers referenced in [32] did not use any quantitative evaluation criterion. In the last years, a few formal mathematical evaluation measures of dimensionality reduction have been proposed in the literature. We argue that one of these measures, the co-ranking framework proposed in [14,16], is directly suitable as a highly flexible and generic tool to evaluate the preservation of pairwise relationships in different dissimilarity measures.

In this section, we give a short overview about the co-ranking framework. Assume points $\mathbf{x}_i$ are mapped to projections $\mathbf{y}_i$ using some dimensionality reduction technique. The co-ranking framework essentially evaluates, in how far neighborhoods in the original space and the projection space correspond to each other. Let δ_{ij} be the distance of $\mathbf{x}_i$ and $\mathbf{x}_j$ and d_{ij} be the distance of $\mathbf{y}_i$ and $\mathbf{y}_j$. The rank of $\mathbf{x}_j$ with respect to $\mathbf{x}_i$ is given by $\rho_{ij} = |\{k|\delta_{ik} < \delta_{ij} \text{ or } (\delta_{ik} = \delta_{ij} \text{ and } k < j)\}|$. Analogously, the rank of r_{ij} for the projections can be defined based on d_{ij}. The co-ranking matrix Q [14] is defined by $Q_{kl} = |\{(i,j)|\rho_{ij} = k \text{ and } r_{ij} = l\}|$. Errors of a dimensionality reduction correspond to rank errors, i.e. off-diagonal entries in this matrix. Usually, the focus of dimensionality reduction is on the preservation of local relationships. In [14], an intuitive measure of rank-preservation has been proposed, the *Quality*

$$Q_{\mathrm{NX}}(K) = \frac{1}{KN} \sum_{k=1}^{K} \sum_{l=1}^{K} Q_{kl}.$$

where N denotes the number of points. This summarizes all 'benevolent' points which change their rank only within a fixed neighborhood K. Essentially, it is the average ratio of all points which stay in a K-neighborhood in the original and the projection space. To get an overall impression of the quality in different neighborhood regimes, usually a curve is plotted for a all possible K or a range thereof. A qualitatively good visualization w.r.t. all K-neighborhoods corresponds to the value $Q_{\mathrm{NX}}(K)$ approaching 1. Interestingly, this framework can be linked to an information theoretic point of view as specified in [33] and it subsumes several previous evaluation criteria, see [14,20]. It is possible to extend this framework to a point-wise evaluation as introduced in [20]. Here, all neighborhood sizes are considered for one fixed point $\mathbf{x}_i$ only, leading to the local quality curve $Q_{\mathrm{NX}}^{\mathbf{x}_i}(K) = \frac{1}{KN} \sum_{k=1}^{K} \sum_{l=1}^{K} Q_{kl}(\mathbf{x}_i)$. Obviously, $Q_{\mathrm{NX}}(K) = \sum_{\mathbf{x}_i} Q_{\mathrm{NX}}^{\mathbf{x}_i}(K)$.

How can this technique be used to compare two dissimilarities? Since $Q_{\mathrm{NX}}(K)$ essentially evaluates in how far a rank-neighborhood induced by δ_{ij} coincides with a rank-neighborhood induced by d_{ij}, we can directly apply this measurement to two given dissimilarity measures d^1 and d^2, and obtain a quantitative statement about the rank-preservation of d^2 given d^1. Since $Q_{\mathrm{NX}}(K)$ is symmetric, the ordering of the dissimilarities is not important.

4   Comparison of Metrics for the Euclidean Vector Space

We start with an illustrative example which shows that the measure $Q_{\mathrm{NX}}(K)$ allows to identify situations where dissimilarities induce similar/dissimilar results. We restrict to the two-dimensional Euclidean vector space where data are distributed uniformly or in clustered form, respectively, see Fig. 1. For these data, we compare the Euclidean distance to the L_k norm, with $k \in \{1,3,6\}$ as well as the maximum-norm as the limit case. We can see the effect of these choices by using a metric multidimensional scaling (MDS) to project the data to the Euclidean plane, see Figs. 2 and 3. Obviously, if data is distributed uniformly,

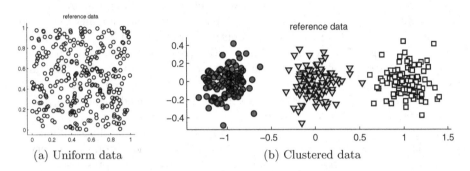

Fig. 1. Original data in the two-dimensional plane with uniform distribution (a) or clustered distribution (b).

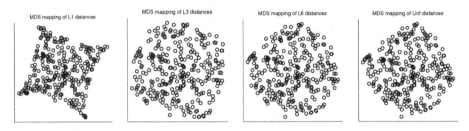

Fig. 2. Comparison of L_k-norms on uniform square data. (L_1, L_3, L_6, L_∞ l.t.r.)

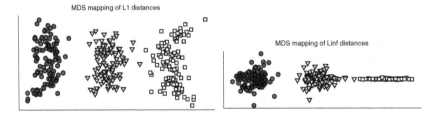

Fig. 3. MDS projection using L_p-norms on three clusters data. (L_1, L_∞)

a smooth transition from L_1 to L_∞ can be observed, as expected, whereby the global topological form does not change much. This observation is mirrored in the co-ranking evaluation, see Fig. 4. The quality curves change smoothly and have a value near 1, indicating a good agreement of the topologies. Note that these metrics are topologically equivalent in the mathematical sense, which is supported by the observation made in this case.

The situation changes if more realistic settings are considered, i.e. if structure is present in the data. We consider three clusters and the same setting as before. Here, the metric L_1 and L_∞ yield very different behavior, as can be seen in the projection in Fig. 3 as well as in the evaluation in Fig. 4. Thus, mathematical topology equivalence does not imply that the overall topologies are similar for

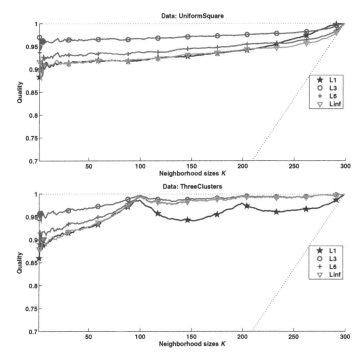

Fig. 4. Comparison of the dissimilarities using the co-ranking framework: uniform square (top) and three clusters (bottom)

realistic settings displaying structure. The co-ranking framework mirrors the expected differences in these settings. Note, that due to the choice of K, also differences at different scales are displayed. In Fig. 4, clearly the underlying structure with cluster sizes of 100 can be recovered from the quality curves.

5   Comparison of Non-euclidean Settings

In the previous sections, we introduced a mathematical approach to compare two dissimilarity measures, and demonstrated it on artificial data sets. In this section, we use two real world data scenarios as a first proof-of-concept study, to show the usefulness of our approach given domain-specific – and possibly non-Euclidean – dissimilarity measures.

App Description Texts. Current research in the area of semantic web utilizes state-of-the-art machine learning and data visualization techniques, in order to automatically organize and represent vast data collections within user-friendly interfaces. Here, sophisticated data dissimilarity measures for textual content play an important role. Our first experimental scenario relates to a typical machine learning task in this context. It consists of descriptions from 500 randomly collected applications, available on the online platform *Google Play* (http://play.google.com). Google Play is a large distribution service for digital multimedia content which currently offers about 450.000 downloadable programs (commonly referred to as *apps*) for the mobile operating system *Android*. Each app is attributed to one of 34 categories, while every category belongs to one of the two major branches "Games" or "Applications". The content of every app is summarized in a textual description of about 1200 characters on average. Our 500 apps come from two categories: 293 from "Arcade & Action" (in Games), and 207 from "Travel & Local" (in Applications). In the following they will be referred to as class 1 and 2, respectively. We consider three different measures to calculate dissimilarities between the descriptions:

(I) *Euclidean* distances on the tf-idf weights, where weight vectors are calculated from the frequencies of the appearing terms (tf) and their inverse frequency of occurrence in all documents (idf), see [25],

(II) the *Cosine* distance on the term frequencies, which is calculated as $c(\mathbf{a}, \mathbf{b}) := 1 - \left((\mathbf{a}^\mathsf{T}\mathbf{b})/(\pi \|\mathbf{a}\| \|\mathbf{b}\|)\right)$, where $\mathbf{a}$ and $\mathbf{b}$ are vectors of term frequencies for the two respective documents,

(III) the *normalized compression distance* (NCD), which is a string dissimilarity measure based on the Kolmogorov complexity [4], in our case using the Lempel-Ziv-Markov chain compressor (LZMA).

While the first two measures are based on basic word statistics, the NCD also takes structural aspects into account implicitly, since the lossless compressor utilizes recurring patterns in the texts to reduce the description length. Prior to applying the dissimilarity measures, we used a standard preprocessing workflow of stopword reduction and Porter stemming.

Fig. 5 shows MDS visualizations of the three different dissimilarities, as well as evaluation curves from the comparison of Euclidean distances versus the Cosine and the NCD measure. For the visualizations in Fig. 5a, 5b, 5d, we used non-metric MDS with squared stress. From the evaluation curves in Fig. 5c we see that the agreement of the Euclidean distances to the Cosine and NCD measure is low in general, with values below 0.6, even for very small neighborhood sizes. Although the visualizations indicate a qualitatively similar structure, the overall ranks seem to be rather different, which is also reflected in the visualizations to some extent: Fig. 5a shows a small number of outliers, while there are fairly distinct clusters in Fig. 5d; and Fig. 5b shows both characteristics: similarly dense regions and some widespread outliers. In this real world data set, every pair of measures showed a low agreement when compared with the evaluation framework, with $Q_{\mathrm{NX}}(K) < 0.6$ for all $K < 100$.

Java Programs. The second example is related to current challenges in the research of *intelligent tutoring systems* (ITS). In general, these educational technology systems are intended to provide intelligent, one-on-one, computer-based support to students in various learning scenarios. Especially in situations where this type of learning support is not available due to scarce (human) resources, the benefits of ITSs become apparent. Since traditional ITSs rely on an exact formalization of the underlying domain knowledge in order to judge whether a given answer from a student is correct or not, they are today mainly applied in well-structured and comparably narrow domains. In order to make future ITSs more flexible, current approaches suggest the application of machine learning techniques to automatically infer models from given sets of student solutions, see [28]. The structural aspects of such data is hard to represent in vectors of numerical features, which would yield an embedding in a Euclidean vector space. Instead, a crucial ingredient of such approaches are domain-specific, and possibly non-metric dissimilarity measures, by which the data can be represented in terms of pairwise relations only. The analysis and development of dissimilarity measures in this area makes a framework for quantitative comparison necessary.

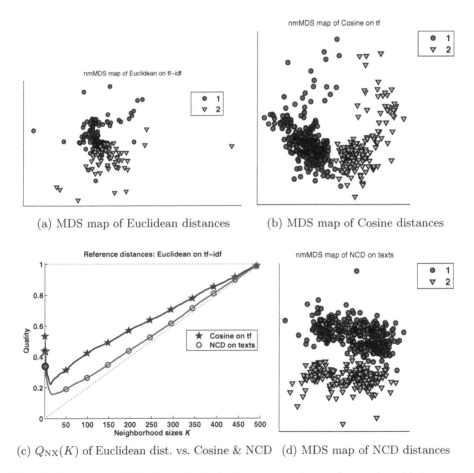

(a) MDS map of Euclidean distances   (b) MDS map of Cosine distances

(c) $Q_{\mathrm{NX}}(K)$ of Euclidean dist. vs. Cosine & NCD   (d) MDS map of NCD distances

Fig. 5. Comparison of the three dissimilarity measures in our first real world showcase scenario consisting of 500 textual descriptions of Android apps

Our data scenario is related to this domain and consists of 169 short Java programs which represent student solutions, originating from a Java programming class of first year students at Clausthal University of Technology, Germany. We used the open source plagiarism detection software *Plaggie* [29] to extract a tokenized representation (a *token stream*) from each given Java source code. Based on the token streams, we consider four different dissimilarity measures:

(I) Euclidean distances on the tf-idf weights like in the previous data set, however, tf and idf now refer to the occurrence of each token instead of term,
(II) the Cosine distance on the token frequencies,
(III) the normalized compression distance (NCD) on the token streams,
(IV) *Greedy String Tiling* (GST) which is the inherent similarity measure that Plaggie uses to compare the given sources [29,30]; since GST yields a matrix S of pairwise similarities $s(\mathbf{x}^i, \mathbf{x}^j) \in S$, where values are in $(0, 1)$ and self-similarities equal 1, we converted S into a dissimilarity matrix by taking $D := \sqrt{1 - S}$, as proposed in [23].

Fig. 6 shows the quality $Q_{\mathrm{NX}}(K)$ when comparing Euclidean distances to Cosine, GST, and NCD dissimilarities. The curves show the highest similarity to the Cosine distances, especially high in small neighborhood ranges, which is expected due to the fact that both are based on token frequencies. Interestingly, the curves of the Cosine and the GST measure show a similar shape in comparison to Euclidean distances, which may indicate a similar response to certain structural aspects in the data, in contrast to the steadily growing curve for NCD.

Fig. 7 demonstrates our proposed framework for the pointwise comparison of dissimilarity measures on the same data scenario. The coloring in 7c and 7d refers to $Q_{\mathrm{NX}}^{\mathbf{x}_i}(20)$, which is the agreement of the 20-neighborhood for every point $\mathbf{x}_i$ as compared to the other dissimilarity measure. To link the coloring scheme to the evaluation curves, $K = 20$ is highlighted on the graphs in Fig. 6. The pointwise evaluation clearly reveals a region of data which is very close in the Euclidean case, but was considered very dissimilar by the GST measure.

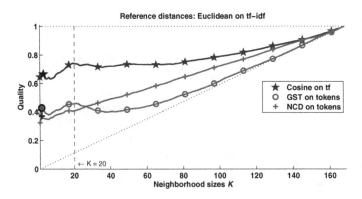

Fig. 6. $Q_{\mathrm{NX}}(K)$ when comparing Euclidean distances to Cosine, GST, and NCD dissimilarities used on our second showcase data set consisting of 169 student solutions from a Java programming class

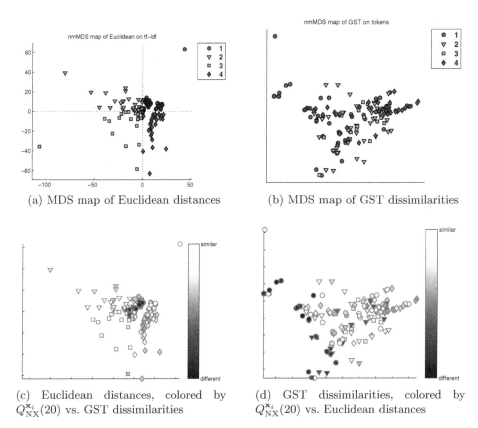

(a) MDS map of Euclidean distances   (b) MDS map of GST dissimilarities

(c) Euclidean distances, colored by $Q_{\mathrm{NX}}^{\mathbf{x}_i}(20)$ vs. GST dissimilarities   (d) GST dissimilarities, colored by $Q_{\mathrm{NX}}^{\mathbf{x}_i}(20)$ vs. Euclidean distances

Fig. 7. Pointwise comparison of dissimilarity measures used on a data set of 169 student solutions from a Java programming class. The dissimilarities from two measures (Euclidean and GST) are mapped to 2D using non-metric MDS. The different symbols for points in the visualizations do not correspond to semantical data classes, but to the quadrants of the cartesian coordinate system in (a), to give some indication of how the point locations differ to the map of GST in (b). The pointwise coloring in (c) and (d) shows for each point, how much the neighbor ranks in the Euclidean case differ to the ranks given by GST.

6   Discussion

We have discussed possibilities to compare dissimilarity measures for unsupervised machine learning tasks. We argued, that rank-preservation or, alternatively, an information retrieval perspective seem very suitable and can be formalized by means of the co-ranking framework taken from the evaluation of dimensionality reduction. We have demonstrated the usefulness in one illustrative artificial example referring to Euclidean vector spaces, as well as two real world examples with problem-specific metrics. The results show that this proposal offers a promising step towards the evaluation, in how far different dissimilarity measures or different choices of metric parameters can lead to substantially different results, when used for unsupervised machine learning.

Naturally, further evaluation techniques are possible such as an evaluation based on the mutual information of the dissimilarities, for example. We conjecture, however, that an information theoretic perspective leads to results which are similar to the co-ranking framework. This is the subject of ongoing work. Further, it is necessary to test whether this a priori comparison of dissimilarity measures coincides with their behavior in typical unsupervised machine learning tasks. Actually, we have already evaluated this behavior to some extent, when visualizing the data in this contribution. The test of further visualization and clustering techniques will be the subject of future work.

Acknowledgment. This work has been supported by the German Science Foundation (DFG) under grants number PI764/6 and HA2719/6-1, and by the center of excellence for cognitive interaction technology (CITEC).

References

1. Ackerman, M., Ben-David, S., Loker, D.: Towards property-based classification of clustering paradigms. In: NIPS 2010, pp. 10–18 (2010)
2. Cha, S.-H.: Comprehensive survey on distance/similarity measures between probability density functions. Int. J. of Mathematical Models and Methods in Appl. Sci. 1(4), 300–307 (2007)
3. Chen, Y., Garcia, E.K., Gupta, M.R., Rahimi, A., Cazzanti, L.: Similarity-based classification: Concepts and algorithms. JMLR 10, 747–776 (2009)
4. Cilibrasi, R., Vitányi, P.: Clustering by compression. IEEE Trans. on Information Theory 51(4), 1523–1545 (2005)
5. Frasconi, P., Gori, M., Sperduti, A.: A general framework for adaptive processing of data structures. IEEE TNN 9(5), 768–786 (1998)
6. Gärtner, T.: Kernels for Structured Data. PhD thesis, Univ. Bonn (2005)
7. Gisbrecht, A., Mokbel, B., Hammer, B.: Relational generative topographic mapping. Neurocomputing 74(9), 1359–1371 (2011)
8. Hammer, B., Hasenfuss, A.: Topographic mapping of large dissimilarity datasets. Neural Computation 22(9), 2229–2284 (2010)
9. Hammer, B., Jain, B.: Neural methods for non-standard data. In: ESANN 2004, pp. 281–292 (2004)
10. Hammer, B., Micheli, A., Sperduti, A.: Universal approximation capability of cascade correlation for structures. Neural Computation 17, 1109–1159 (2005)
11. Hammer, B., Micheli, A., Sperduti, A.: Adaptive Contextual Processing of Structured Data by Recursive Neural Networks: A Survey of Computational Properties. In: Hammer, B., Hitzler, P. (eds.) Perspectives of Neural-Symbolic Integration. SCI, vol. 77, pp. 67–94. Springer, Heidelberg (2007)
12. Hammer, B., Mokbel, B., Schleif, F.-M., Zhu, X.: White Box Classification of Dissimilarity Data. In: Corchado, E., Snášel, V., Abraham, A., Woźniak, M., Graña, M., Cho, S.-B. (eds.) HAIS 2012, Part III. LNCS, vol. 7208, pp. 309–321. Springer, Heidelberg (2012)
13. Hathaway, R.J., Bezdek, J.C.: Nerf c-means: Non-euclidean relational fuzzy clustering. Pattern Recognition 27(3), 429–437 (1994)

14. Lee, J.A., Verleysen, M.: Quality assessment of dimensionality reduction: Rank-based criteria. Neurocomputing 72(7-9), 1431–1443 (2009)
15. Lee, J.A., Verleysen, M.: Nonlinear dimensionality redcution. Springer (2007)
16. Lee, J.A., Verleysen, M.: Scale-independent quality criteria for dimensionality reduction. Pattern Recognition Letters 31, 2248–2257 (2010)
17. Lewis, J., Ackerman, M., Sa, V.D.: Human cluster evaluation and formal quality measures. In: Proc. of the 34th Ann. Conf. of the Cog. Sci. Society (2012)
18. Liu, H., Song, D., Rüger, S., Hu, R., Uren, V.: Comparing Dissimilarity Measures for Content-Based Image Retrieval. In: Li, H., Liu, T., Ma, W.-Y., Sakai, T., Wong, K.-F., Zhou, G. (eds.) AIRS 2008. LNCS, vol. 4993, pp. 44–50. Springer, Heidelberg (2008)
19. Malerba, D., Esposito, F., Gioviale, V., Tamma, V.: Comparing dissimilarity measures for symbolic data analysis. In: Pre-Proc. of ETK-NTTS 2001, HERSONIS-SOS, pp. 473–481 (2001)
20. Mokbel, B., Lueks, W., Gisbrecht, A., Biehl, M., Hammer, B.: Visualizing the quality of dimensionality reduction. In: ESANN 2012, pp. 179–184 (2012)
21. Neuhaus, M., Bunke, H.: Edit distance-based kernel functions for structural pattern classification. Pat. Rec. 39(10), 1852–1863 (2006)
22. Pearson, W.R., Lipman, D.J.: Improved tools for biological sequence comparison. Proc. of the National Academy of Sciences USA 85(8), 2444–2448 (1988)
23. Pekalska, E., Duin, R.P.: The Dissimilarity Representation for Pattern Recognition. Foundations and Applications. World Scientific (2005)
24. Qin, A.K., Suganthan, P.N.: Kernel neural gas algorithms with application to cluster analysis. In: ICPR 2004, vol. 4, pp. 617–620. IEEE Computer Society (2004)
25. Robertson, S.: Understanding inverse document frequency: On theoretical arguments for idf. Journal of Documentation 60(5), 503–520 (2004)
26. Rossi, F., Villa-Vialaneix, N.: Consistency of functional learning methods based on derivatives. Pat. Rec. Letters 32(8), 1197–1209 (2011)
27. Scarselli, F., Gori, M., Tsoi, A.C., Hagenbuchner, M., Monfardini, G.: Computational capabilities of graph neural networks. IEEE TNN 20(1), 81–102 (2009)
28. Gross, S., Zhu, X., Hammer, B., Pinkwart, N.: Cluster Based Feedback Provision Strategies in Intelligent Tutoring Systems. In: Cerri, S.A., Clancey, W.J., Papadourakis, G., Panourgia, K. (eds.) ITS 2012. LNCS, vol. 7315, pp. 699–700. Springer, Heidelberg (2012)
29. Mozgovoy, M., Karakovskiy, S., Klyuev, V.: Fast and reliable plagiarism detection system. In: 37th Annual Frontiers In Education Conference - Global Engineering: Knowledge Without Borders, Opportunities Without Passports, FIE 2007 (2007)
30. Wise, M.J.: Running Karp-Rabin Matching and Greedy String Tiling. Technical report 463 (Univ. of Sydney. Basser Dept. of Comp. Sci.) (1993) ISBN 0867586699
31. van der Maaten, L., Hinton, G.: Visualizing high-dimensional data using t-sne. JMLR 9, 2579–2605 (2008)
32. Venna, J.: Dimensionality reduction for Visual Exploration of Similarity Structures. PhD thesis, Helsinki University of Technology, Espoo, Finland (2007)
33. Venna, J., Peltonen, J., Nybo, K., Aidos, H., Kaski, S.: Information retrieval perspective to nonlinear dimensionality reduction for data visualization. JMLR 11, 451–490 (2010)
34. Yin, H.: On the equivalence between kernel self-organising maps and self-organising mixture density networks. Neural Netw. 19(6), 780–784 (2006)

Kernel Robust Soft Learning Vector Quantization

Daniela Hofmann and Barbara Hammer

CITEC Center of Excellence, Bielefeld University, Germany
{dhofmann,bhammer}@techfak.uni-bielefeld.de

Abstract. Prototype-based classification schemes offer very intuitive and flexible classifiers with the benefit of easy interpretability of the results and scalability of the model complexity. Recent prototype-based models such as robust soft learning vector quantization (RSLVQ) have the benefit of a solid mathematical foundation of the learning rule and decision boundaries in terms of probabilistic models and corresponding likelihood optimization. In its original form, they can be used for standard Euclidean vectors only. In this contribution, we extend RSLVQ towards a kernelized version which can be used for any positive semidefinite data matrix. We demonstrate the superior performance of the technique, kernel RSLVQ, in a variety of benchmarks where results competitive or even superior to state-of-the-art support vector machines are obtained.

1 Introduction

A variety of powerful classification, regression, and inference techniques being available, machine learning has revolutionized the possibility to deal with large electronic data sets and to infer models for complex settings where standard statistical models are no longer sufficient. Because of its high flexibility and its usually excellent classification and generalization performance, the support vector machine (SVM) constitutes one of the current flagships of supervised machine learning. With machine learning techniques becoming more and more popular in diverse application domains, there is an increasing need for models which can easily be used by applicants outside the field of machine learning or computer science. Moreover, due to more and more complex data and settings, the tasks become more and more complex and, often, applicants do not only have to apply a machine learning technique but also to inspect and interpret the result. Based on insight gained this way, an improvement or focus of the model can be done [23]. In this setting, a severe drawback of many state-of-the-art machine learning tools occurs: they act as black-boxes. In consequence, applicants cannot interpret the results and it is hardly possible to substantiate a machine classification by a semantic explanation, or to change the functionality of the model based on this insight.

Prototype-based methods enjoy a wide popularity in various application domains due to their very intuitive and simple behavior: they represent their decisions in terms of typical representatives contained in the input space and a

N. Mana, F. Schwenker, and E. Trentin (Eds.): ANNPR 2012, LNAI 7477, pp. 14–23, 2012.

classification is based on the distance of data as compared to these prototypes [12]. Thus, models can be directly inspected by experts since prototypes can be treated in the same way as data. Popular techniques in this context include simple learning vector quantization (LVQ) schemes and extensions to more powerful settings such as variants based on cost functions or metric learners [18,21]. Robust soft LVQ (RSLVQ) as proposed in [21] constitutes one particularly interesting approach since it is based on a generic probabilistic modeling of data in terms of mixture models and it derives a learning rule based on this model by optimizing the likelihood ratio. A behavior which closely resembles standard LVQ2.1 results if modes are represented as Gaussians and the limit case of small bandwidth is considered. While the limit case as well as standard LVQ2.1 do not achieve optimum behavior already in simple model situations, as investigated in the context of the theory of online learning in the approach [1] for example, RSLVQ displays excellent generalization ability in the standard intermediate case, see e.g. the approach [20] for an extensive comparison of the technique.

With data sets becoming more and more complex, input data are often no longer given as simple Euclidean feature vectors, rather structured data or dedicated formats can be observed such as bioinformatics sequences, graphs, or tree structures as they occur in linguistics, time series data, functional data arising in mass spectrometry, relational data stored in relational databases, etc. In consequence, a variety of techniques has been developed to extend powerful statistical machine learning tools towards non-vectorial data such as kernel methods using structure kernels, recursive and graph networks, functional methods, relational approaches, and similar [6,19,8,17,10]. Recently, popular prototype-based algorithms have also been extended to deal with more general data. Diverse techniques rely on a characterization of the data by means of a matrix of pairwise similarities or dissimilarities only rather than explicit feature vectors. In this setting, median clustering as provided by median self-organizing maps, median neural gas, or affinity propagation characterizes clusters in terms of typical exemplars [7,13,5]. More general smooth adaptation is offered by relational extensions such as relational neural gas or relational learning vector quantization [9]. A further possibility is offered by kernelization such as proposed for neural gas, self-organizing maps, or different variants of learning vector quantization [15,3,16]. By formalizing the interface to the data as a general similarity or dissimilarity matrix, complex structures can be easily dealt with: structure kernels for graphs, trees, alignment distances, string distances, etc. open the way towards these general data structures [14,8].

In this contribution, we propose an extension of RSLVQ towards a kernel variant. This way, a statistically well motivated model is obtained which achieves excellent results as we will show in several benchmarks. Interestingly, albeit the method, strictly speaking, requires a semi positive definite kernel, it also yields good results if applied to arbitrary dissimilarity matrices. Corrections which turn the latter towards valid kernels can further improve the results. Now we first shortly review RSLVQ and we explain how this technique can be extended to a kernelized version. We evaluate the behavior for several benchmarks and also show first visualizations which emphasize the interpretability of the resulting models in terms of prototypes. We conclude with a discussion.

2 Robust Soft Learning Vector Quantization

Learning vector quantization (LVQ) constitutes a very popular class of intuitive prototype based learning algorithms with successful applications ranging from telecommunications to robotics [12]. Basic algorithms as proposed by Kohonen include LVQ1 which is directly based on Hebbian learning, and improvements such as LVQ2.1, LVQ3, or OLVQ which aim at a higher convergence speed or better approximation of the Bayesian borders. These types of LVQ schemes have in common that their learning rule is essentially heuristically motivated and a valid cost function does not exist [2]. One of the first proposals which derives LVQ from a cost function can be found in [18] with an exact computation of the validity at class boundaries in [11]. One very elegant LVQ scheme which is based on a probabilistic model and which can be seen as a more robust probabilistic extension of LVQ2.1 has been proposed in [21]. This method, robust soft LVQ (RSLVQ) models data by means of a mixture of Gaussians and derives learning rules thereof by means of a maximization of the log likelihood ratio of the given data. In the limit of small bandwidth, a learning rule which is similar to LVQ2.1 but which performs adaptation in case of misclassification only, is obtained.

Assume data $\xi_k \in \mathbb{R}^n$ are labeled y_k where labels stem from a finite number of different classes. A RSLVQ network models data by means of a mixture distribution characterized by m prototypes $w_j \in \mathbb{R}^n$ with priorly fixed labels $c(w_j)$ and bandwidths σ_j. Mixture component j defines the probability

$$p(\xi|j) = K(j) \cdot \exp(f(\xi, w_j, \sigma_j^2))$$

with normalization constant $K(j)$ and function f chosen e.g. as follows

$$f(\xi, w_j, \sigma_j^2) = -\|\xi - w_j\|^2 / \sigma_j^2$$

based on the Euclidean distance or a generalization thereof. This induces the probability of an unlabeled data point

$$p(\xi|W) = \sum_j P(j) \cdot p(\xi|j)$$

with prior $P(j)$ and parameters W of the model. The probability of a labeled data point is

$$p(\xi, y|W) = \sum_{c(w_j)=y} P(j) \cdot p(\xi|j).$$

Learning aims at an optimization of the log likelihood ratio

$$L = \sum_k \log \frac{p(\xi_k, y_k|W)}{p(\xi_k|W)}.$$

A stochastic gradient ascent yields the following update rules, given data point (ξ_k, y_k)

$$\Delta w_j = \alpha \cdot \begin{cases} (P_y(j|\xi_k) - P(j|\xi_k)) \cdot K(j) \cdot \partial f(\xi_k, w_j, \sigma_j^2)/\partial w_j & \text{if } c(w_j) = y_k \\ -P(j|\xi_k) \cdot K(j) \cdot \partial f(\xi_k, w_j, \sigma_j^2)/\partial w_j & \text{if } c(w_j) \neq y_k \end{cases}$$

with learning rate $\alpha > 0$ and the probabilities

$$P_y(j|\xi_k) = \frac{P(j) \exp(f(\xi_k, w_j, \sigma_j^2))}{\sum_{c(w_j)=y_j} P(j) \exp(f(\xi_k, w_j, \sigma_j^2))}$$

and

$$P(j|\xi_k) = \frac{P(j) \exp(f(\xi_k, w_j, \sigma_j^2))}{\sum_j P(j) \exp(f(\xi_k, w_j, \sigma_j^2))}$$

With the standard Euclidean distance, equal class priors, and small bandwidth, a learning rule similar to LVQ2.1, learning from mistakes, results thereof.

Given a novel data point ξ, its class label can be determined by means of the most likely label y corresponding to a maximum value $p(y|\xi, W) \sim p(\xi, y|W)$. For typical settings, bandwidths are chosen of equal size $\sigma_j^2 = \sigma^2$, and priors are equal $P(j) = \text{const}$. Further, the simple Euclidean distance is used. Then, this rule can usually be approximated by a simple winner takes all rule, i.e. ξ is mapped to the label $c(w_j)$ of the closest prototype w_j. It has been shown in [21], for example, that RSLVQ often yields excellent results while preserving interpretability of the model due to prototypical representatives of the classes in terms of the parameters w_j.

3   Kernel Robust Soft Learning Vector Quantization

RSLVQ, albeit offering a very powerful learning algorithm, is restricted to Euclidean data only. Here we propose a kernelization of the method such that the technique becomes applicable for more general data sets which are implicitly characterized in terms of a Gram matrix only. We assume that a kernel k is fixed corresponding to a feature map Φ, hence

$$k_{kl} := k(\xi_k, \xi_l) = \Phi(\xi_k)^t \Phi(\xi_l)$$

holds for all data points ξ_k, ξ_l. We assume that prototypes are represented as linear combinations of data in the feature space

$$w_j = \sum_m \gamma_{jm} \Phi(\xi_m).$$

It is reasonable to assume that they are contained in the convex hull of the data, i.e. coefficients γ_{jm} are non-negative and sum up to 1. The cost function of RSLVQ becomes

$$L = \sum_k \log \frac{\sum_{c(w_j)=y_k} P(j)p(\Phi(\xi_k)|j)}{\sum_j P(j)p(\Phi(\xi_k)|j)}.$$

We assume equal bandwidth $\sigma^2 = \sigma_j^2$, constant prior $P(j)$ and mixture components induced by normalized Gaussians. These can be computed in the data space based on the Gram matrix because of the identity

$$\|\Phi(\xi_i) - w_j\|^2 = \|\Phi(\xi_i) - \sum_m \gamma_{jm}\Phi(\xi_m)\|^2 = k_{ii} - 2 \cdot \sum_m \gamma_{jm}k_{im} + \sum_{s,t} \gamma_{js}\gamma_{jt}k_{st}$$

where the distance in the feature space is referred to by $\| \cdot \|^2$. Thus the update rules become $\Delta w_j = \sum_m \Delta\gamma_{jm}\Phi(\xi_m) =$

$$\alpha \cdot K(j) \cdot \begin{cases} (P_y(j|\Phi(\xi_k)) - P(j|\Phi(\xi_k)))\,(\Phi(\xi_k) - \sum_m \gamma_{jm}\Phi(\xi_m)) & \text{if } c(w_j) = y_k \\ -P(j|\Phi(\xi_k))\,(\Phi(\xi_k) - \sum_m \gamma_{jm}\Phi(\xi_m)) & \text{if } c(w_j) \neq y_k \end{cases}$$

Hence a gradient technique yields the following adaptation rules for the coefficients γ_{jm}:

$$\Delta\gamma_{jm} = \alpha \cdot K(j) \cdot \begin{cases} -(P_y(j|\Phi(\xi_k)) - P(j|\Phi(\xi_k)))\gamma_{jm} & \text{if } \xi_m \neq \xi_k, c(w_j) = y_k \\ (P_y(j|\Phi(\xi_k)) - P(j|\Phi(\xi_k)))(1 - \gamma_{jm}) & \text{if } \xi_m = \xi_k, c(w_j) = y_k \\ P(j|\Phi(\xi_k))\gamma_{jm} & \text{if } \xi_m \neq \xi_k, c(w_j) \neq y_k \\ -P(j|\Phi(\xi_k))(1 - \gamma_{jm}) & \text{if } \xi_m = \xi_k, c(w_j) \neq y_k \end{cases}$$

Note that this adaptation performs exactly the same updates as RSLVQ in the feature space provided that the prototypes can be expressed as linear combinations of data points in the feature space. To guarantee non-negative and normalized coefficients, simple normalization takes place after every adaptation step. This restriction to the convex hull of the feature space is reasonable: it has been demonstrated e.g. in [20] that RSLVQ, by learning from mistakes, does not necessarily place prototypes at typical positions of the data space if this does not further improve the classification accuracy, rather orthogonal transformations are accepted in this case, leading to unintuitive representations of the data. These ambiguities of the solution are avoided by referring to the convex hull.

4   Experiments

We compare the method to the support vector machine (SVM) and a k-nearest neighbor classifier (k-NN) on a variety of benchmarks as introduced in [4]. The data sets represent a variety of similarity matrices which are, in general, non-Euclidean. It is standard to symmetrize the matrices by taking the average of the matrix and its transposed. Further, the substitution of a given similarity by its normalized variant constitutes a standard preprocessing step, arriving at diagonal entries 1. Even in symmetrized and normalized form, the matrices do not necessarily provide a valid kernel. Hence k-NN is directly applicable, while SVM and, strictly speaking, kernel RSLVQ, are not. We observe, however, that, unlike SVM, kernel RSLVQ can deal with these data directly without any correction due to its direct optimization of the cost function by means of a gradient descent method.

There exist different standard preprocessing tools which transfer a given similarity matrix into a valid kernel, as presented e.g. in [4,14]. In general, the similarity matrix can posses negative eigenvalues which yield to an invalid kernel. Corrections are:

- *Spectrum clip:* simply set negative eigenvalues of the matrix to 0. Since this can be realized as a linear projection, it directly transfers to out-of-sample extensions.
- *Spectrum flip:* negative eigenvalues are substituted by their positive values. Again, this can be realized by means of a linear transformation.
- *Spectrum shift:* the absolute value of the smallest negative eigenvalue is added to all eigenvalues. For spectrum shift there does not exist an according linear transform. Since the transform only affects self-similarities, a possible out-of-sample extension is to let the new similarities unchanged.

These transforms are tested for kernel RSLVQ in comparison to SVM with according preprocessing and a k-nearest neighbor approach with kernel ridge regression weights. For the latter, we report results taken from [4]. We use training data in analogy to [4]. For all data sets, we also report the signature, i.e. the number of positive and negative eigenvalues of the Gram matrix, indicating the degree of non-Euclideanity of the data.

- *Amazon47*: This data set consists of 204 books written by four different authors. The similarity is determined as the percentage of customers who purchase book j after looking at book i. This matrix is fairly sparse and mildly non-Euclidean with signature $(191, 13, 0)$. Class labeling of a book is given by the author.
- *Aural Sonar*: This data set consists of 100 wide band solar signals corresponding to two classes, observations of interest versus clutter. Similarities are determined based on human perception, averaging over 5 random probands for each signal pair. The signature is $(62, 38, 0)$. Class labeling is given by the two classes: target of interest versus clutter.
- *Face Rec*: 945 images of faces of 139 different persons are recorded. Images are compared using the cosine-distance of integral invariant signatures based on surface curves of the 3D faces. The signature is given by $(794, 151, 0)$. The labeling corresponds to the 139 different persons.
- *Patrol*: 241 samples representing persons in seven different patrol units are contained in this data set. Similarities are based on responses of persons in the units about other members of their groups. The signature is $(116, 125, 0)$. Class labeling corresponds to the seven patrol units.
- *Protein*: 213 proteins are compared based on evolutionary distances comprising four different classes according to different globin families. The signature is $(171, 72, 0)$. Labeling is given by four classes corresponding to globin families.
- *Voting*: Voting contains 435 samples with categorical data compared by means of the value difference metric. Class labeling into two classes is present. The signature is $(225, 210, 0)$.

For these data sets, results for the SVM and a weighted k-NN classifier have been reported in [4]. Thereby, data are preprocessed using shift, clip, or flip to guarantee positive definiteness for SVM. The latter is used with the RBF kernel and optimized meta-parameters in [4]. For multi-class classification, the one versus one scheme has been used.

In comparison, we train a kernel RSLVQ network using the real data or its clip, flip, or shift, respectively. Results of a ten-fold cross-validation with the same partitioning as proposed in [4] are reported. Prototypes are initialized by means of normalized random coefficients γ_{jm} where the prior class label $c(w_j)$ determines the non-zero elements. Further, while training, we guarantee that prototypes are contained in the convex hull of the data by enforcing non-negative coefficients and normalized vectors after every adaptation step. The number of prototypes is taken as a small multiple of the number of classes, exact values being displayed in Tab. 1. Other meta-parameters are optimized on the data sets using cross-validation.

The results obtained on these data sets are reported in Tab. 1, whereby results for k-NN and SVM are taken from [4]. Since SVM requires a positive semidefinite matrix, only results for the corrected data are reported for SVM. For kernel RSLVQ, albeit it is defined for valid kernels only in the strict sense, a direct application for the original data leads to (often very competitive) results which are reported in Tab. 1. For every data set, the best achieved result is shown in boldface. Interestingly, in half the cases, kernel RSLVQ achieves the best result. For four out of six cases, already the performance for the original data beats the SVM result for pre-processed data. Only in two cases (Protein and Voting), kernel RSLVQ is substantially worse as compared to SVM, albeit the result still stays in the same order of magnitude. Overall, it can be inferred that kernel RSLVQ constitutes a very competitive algorithm with excellent classification results overall.

Since prototypes are represented only implicitly by means of coefficient vectors, a direct inspection of a kernel RSLVQ classifier in the same way as a standard LVQ network by inspecting the prototype vectors is not possible. There are two possibilities which still allow an intuitive inspection of the result: since prototypes are contained in the convex hull of the data, it is possible to approximate prototypes by means of the closest data point without too much loss of information. This approximation by exemplars enables its inspection in the same way as data points. As an alternative, pairwise dissimilarities of data and prototypes are given for both, prototypes in its original form as well as exemplar based approximations. Thus it is possible to display data and prototypes in two dimensions by means of a standard non-linear dimensionality reduction technique such as t-SNE which relies on dissimilarities only [22].

To illustrate this possibility, we visualize the Aural Sonar and the Voting data set by means of t-SNE. For both cases, a kernel RSLVQ model is trained using only one prototype per class. The respective closest exemplar is marked in the projection. Fig. 1 displays the results. Obviously, representative discriminative positions are chosen as prototypes which have the potential to offer

Table 1. Results of kernel RSLVQ in comparison to SVM and k-NN on different benchmark data. The test error is reported, the standard deviation is given in parenthesis and best results are shown in boldface.

	k-NN	SVM	kernel RSLVQ	prototypes
Amazon47	16.95 (4.85)	75.98 (7.33)	**15.00** (0.33)	94
clip	17.68 (4.75)	81.34 (4.77)	**14.63** (0.26)	
flip	17.56 (4.91)	84.27 (4.33)	**16.70** (0.33)	
shift	17.68 (4.75)	77.68 (6.14)	**13.78** (0.23)	
Aural Sonar	17.00 (7.65)	14.25 (7.46)	**12.50** (0.48)	10
clip	14.00 (6.82)	13.00 (5.34)	**12.50** (0.48)	
flip	12.75 (6.42)	13.25 (5.31)	**12.00** (0.35)	
shift	13.50 (6.73)	14.00 (5.61)	**13.00** (0.43)	
Face Rec	4.23 (1.43)	3.92 (1.29)	**3.67** (0.02)	139
clip	4.15 (1.32)	4.18 (1.25)	**3.67** (0.02)	
flip	4.15 (1.32)	4.18 (1.32)	**3.65** (0.02)	
shift	4.07 (1.33)	4.15 (1.33)	**3.88** (0.01)	
Patrol	**11.88** (4.42)	40.73 (5.95)	17.29 (0.36)	24
clip	**11.56** (4.54)	38.75 (4.81)	17.91 (0.18)	
flip	**11.67** (4.24)	47.29 (5.90)	18.43 (0.24)	
shift	**13.23** (4.48)	40.83 (5.37)	23.33 (0.30)	
Protein	29.88 (9.96)	**2.67** (2.97)	29.06 (0.27)	20
clip	30.35 (9.71)	**5.35** (4.60)	10.00 (0.26)	
flip	31.28 (9.63)	**1.51** (2.36)	3.13 (0.10)	
shift	30.35 (9.71)	**23.49** (7.31)	34.65 (0.31)	
Voting	5.80 (1.83)	**5.52** (1.77)	9.42 (0.05)	20
clip	5.29 (1.80)	**4.89** (2.05)	9.42 (0.05)	
flip	5.23 (1.80)	**4.94** (2.03)	9.42 (0.05)	
shift	5.29 (1.80)	**5.17** (1.87)	9.42 (0.05)	

Fig. 1. Visualizing the Aural Sonar (left) and Voting data (right) sets together with representative exemplars approximating the prototypes of a kernel RSLVQ classifier using t-SNE

interpretability of the results. Thereby it is vital that, unlike support vectors in SVM, representative positions are chosen as prototypes and its number is fixed a priori.

5 Discussion

We have proposed an extension of RSLVQ to a kernel variant and we have shown that this technique yields excellent results on a variety of benchmarks, reaching the classification accuracy of the SVM in all cases. Thereby, unlike SVM, a representation of the data in terms of representative prototypes is given, and the model can be interpreted as a probabilistic mixture model induced by the prototypes, provided the considered similarity measure is a valid kernel. The latter can be achieved by using e.g. flip or clip. In most cases, also the raw similarity matrix can be used albeit it does not constitute a valid kernel. Since data and classification is based on similarities, standard visualization tools such as t-SNE allow to non-linearly project data onto the plane and to inspect the obtained result. We have demonstrated this opportunity for two simple cases, the visualization of more advanced settings being the subject of ongoing work.

While kernelization greatly enhances the applicability of RSLVQ to complex settings, it has the drawback that it trades linear complexity by quadratic one caused by the quadratic size of the similarity matrix. This makes the technique unsuited if large data sets are dealt with. Popular approximation algorithms include e.g. the Nyström approximation to substitute the full Gram matrix by a low-rank counterpart, or patch processing which processes streaming data consecutively in patches relying on a linear subpart of the full Gram matrix only. See e.g. the publications [24,25] for these techniques and [25] for first successful applications in the context of prototype based methods and LVQ schemes. These techniques can directly be integrated into kernel RSLVQ inducing linear time approximation schemes. It is the subject of future work to evaluate the performance of these approximation schemes.

Acknowledgement. This work has been supported by the DFG under grant number HA2719/7-1 and by the CITEC center of excellence.

References

1. Biehl, M., Ghosh, A., Hammer, B.: Dynamics and generalization ability of LVQ algorithms. Journal of Machine Learning Research 8, 323–360 (2007)
2. Biehl, M., Hammer, B., Verleysen, M., Villmann, T. (eds.): Similarity-Based Clustering. LNCS (LNAI), vol. 5400. Springer, Heidelberg (2009)
3. Boulet, R., Jouve, B., Rossi, F., Villa, N.: Batch kernel SOM and related Laplacian methods for social network analysis. Neurocomputing 71(7-9), 1257–1273 (2008)
4. Chen, Y., Garcia, E.K., Gupta, M.R., Rahimi, A., Cazzanti, L.: Similarity-based classification: Concepts and algorithms. JMLR 10, 747–776 (2009)

5. Cottrell, M., Hammer, B., Hasenfuss, A., Villmann, T.: Batch and median neural gas. Neural Networks 19, 762–771 (2006)
6. Frasconi, P., Gori, M., Sperduti, A.: A general framework for adaptive processing of data structures. IEEE TNN 9(5), 768–786 (1998)
7. Frey, B.J., Dueck, D.: Clustering by passing messages between data points. Science 315, 972–976 (2007)
8. Gärtner, T.: Kernels for Structured Data. PhD thesis, Univ. Bonn (2005)
9. Hammer, B., Hasenfuss, A.: Topographic mapping of large dissimilarity datasets. Neural Computation 22(9), 2229–2284 (2010)
10. Hammer, B., Micheli, A., Sperduti, A.: Universal approximation capability of cascade correlation for structures. Neural Computation 17, 1109–1159 (2005)
11. Hammer, B., Villmann, T.: Generalized relevance learning vector quantization. Neural Networks 15(8-9), 1059–1068 (2002)
12. Kohonen, T.: Self-Oganizing Maps, 3rd edn. Springer (2000)
13. Kohonen, T., Somervuo, P.: How to make large self-organizing maps for nonvectorial data. Neural Networks 15(8-9), 945–952 (2002)
14. Pekalska, E., Duin, R.P.: The Dissimilarity Representation for Pattern Recognition. Foundations and Applications. World Scientific (2005)
15. Qin, A.K., Suganthan, P.N.: Kernel neural gas algorithms with application to cluster analysis. In: Proceedings of the 17th International Conference on Pattern Recognition (ICPR 2004), vol. 4, pp. 617–620. IEEE Computer Society, Washington, DC (2004)
16. Qin, A.K., Suganthan, P.N.: A novel kernel prototype-based learning algorithm. In: Proc. of the 17th International Conference on Pattern Recognition (ICPR 2004), Cambridge, UK (August 2004)
17. Rossi, F., Villa-Vialaneix, N.: Consistency of functional learning methods based on derivatives. Pat. Rec. Letters 32(8), 1197–1209 (2011)
18. Sato, A., Yamada, K.: Generalized Learning Vector Quantization. In: NIPS (1995)
19. Scarselli, F., Gori, M., Tsoi, A.C., Hagenbuchner, M., Monfardini, G.: Computational capabilities of graph neural networks. IEEE TNN 20(1), 81–102 (2009)
20. Schneider, P., Biehl, M., Hammer, B.: Distance learning in discriminative vector quantization. Neural Computation 21, 2942–2969 (2009)
21. Seo, S., Obermayer, K.: Soft learning vector quantization. Neural Comput. 15, 1589–1604 (2003)
22. van der Maaten, L., Hinton, G.: Visualizing high-dimensional data using t-sne. JMLR 9, 2579–2605 (2008)
23. Vellido, A., Martin-Guerroro, J.D., Lisboa, P.: Making machine learning models interpretable. In: ESANN 2012 (2012)
24. Williams, C., Seeger, M.: Using the nyström method to speed up kernel machines. In: Advances in Neural Information Processing Systems 13, pp. 682–688. MIT Press (2001)
25. Zhu, X., Gisbrecht, A., Schleif, F.-M., Hammer, B.: Approximation techniques for clustering dissimilarity data. Neurocomputing 90, 72–84 (2012)

Incremental Learning by Message Passing in Hierarchical Temporal Memory

Davide Maltoni[1] and Erik M. Rehn[2]

[1] Biometric System Laboratory, DEIS - University of Bologna, Italy
davide.maltoni@unibo.it
[2] Bernstein Center for Computational Neuroscience, Berlin, Germany
erik.m.rehn@gmail.com

Abstract. Hierarchical Temporal Memory is a biologically-inspired framework that can be used to learn invariant representations of patterns. Classical HTM learning is mainly unsupervised and once training is completed the network structure is frozen, thus making further training quite critical. In this paper we develop a novel technique for HTM (incremental) supervised learning based on error minimization. We prove that error backpropagation can be naturally and elegantly implemented through native HTM message passing based on Belief Propagation. Our experimental results show that a two stage training composed by unsupervised pre-training + supervised refinement is very effective. This is in line with recent findings on other deep architectures.

Keywords: HTM, Deep architectures, Backpropagation, Incremental learning.

1 Introduction

Hierarchical Temporal Memory (HTM) is a biologically-inspired pattern recognition framework fairly unknown to the research community [1]. It can be conveniently framed into *Multi-stage Hubel-Wiesel Architectures* [2] which is a specific subfamily of Deep Architectures [3-4]. HTM tries to mimic the feed-forward and feedback projections thought to be crucial for cortical computation. Bayesian Belief Propagation is used in a hierarchical network to learn invariant spatio-temporal features of the input data and theories exist to explain how this mathematical model could be mapped onto the cortical-thalamic anatomy [5-6]. A comprehensive description of HTM architecture and learning algorithms is provided in [7], where HTM was also proved to perform well on some pattern recognition tasks, even though further studies and validations are necessary.

One limitation of the classical HTM learning is that once a network is trained it is hard to learn from new patterns without retraining it from scratch. In other words a classical HTM is well suited for a batch training based on a fixed training set, and it cannot be effectively trained incrementally over new patterns that were initially unavailable. In fact, every HTM level has to be trained individually, starting from the bottom: altering the internal node structure at one network level (e.g. coincidences, groups) would invalidate the results of the training at higher levels. In principle, incremental learning could be carried out in a classical HTM by updating only the

N. Mana, F. Schwenker, and E. Trentin (Eds.): ANNPR 2012, LNAI 7477, pp. 24–35, 2012.

output level, but this is a naive strategy that works in practice only if the new incoming patterns are very similar to the existing ones in terms of "building blocks". Since incremental training is a highly desirable property of a learning system, we were motivated to investigate how HTM framework could be extended in this direction.

In this paper we present a two-stage training approach, unsupervised pre-training + supervised refinement, that can be used for incremental learning: a new HTM is initially pre-trained (batch), then its internal structure is incrementally updated as new labeled samples become available. This kind of unsupervised pre-training and supervised refinement was recently demonstrated to be successful for other deep architectures [3]. The basic idea of our approach is to perform the batch pre-training using the algorithms described in [7] and then fix coincidences and groups throughout the whole network; then, during supervised refinement adapt the elements of the probability matrices **PCW** (for the output node) and **PCG** (for the intermediate nodes) as if they were the weights of a MLP neural network trained with backpropagation. To this purpose we derived the update rules based on the descent of the error function. Since the HTM architecture is more complex than MLP, the resulting equations are not simple; further complications arise from the fact that **PCW** and **PCG** values are probabilities and need to be normalized after each update step. Fortunately we found a surprisingly simple (and computationally light) way to implement the whole process through native HTM message passing. Our initial experiments show very promising results. Furthermore, the proposed two-stage approach not only enables incremental learning, but is also helpful for keeping the network complexity under control, thus improving the framework scalability.

2   Background

An HTM has a hierarchical tree structure. The tree is built up by n_{levels} levels (or layers), each composed of one or more nodes. A node in one level is bidirectionally connected to one or more nodes in the level above and the number of nodes in each level decreases as we ascend the hierarchy. The lowest level, $\mathcal{L}_0$, is the input level and the highest level, $\mathcal{L}_{n_{levels}-1}$, with typically only one node, is the output level. Levels and nodes in between input and output are called intermediate levels and nodes. When an HTM is used for visual inference, as is the case in this study, the input level typically has a retinotopic mapping of the input. Each input node is connected to one pixel of the input image and spatially close pixels are connected to spatially close nodes. Refer to Figure 1 of [10] for a graphical example of HTM.

2.1   Information Flow

In an HTM the flow of information is bidirectional. Belief propagation is used to pass messages/information both up (feed-forward) and down (feedback) the hierarchy as new evidence is presented to the network. The notation used here for belief propagation (Figure 1.a) closely follows Pearl [8] and is adapted to HTMs by George [9]:

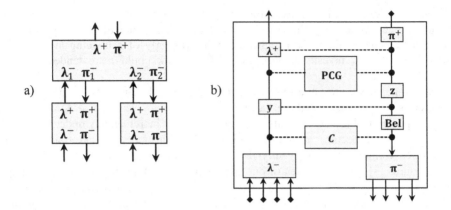

Fig. 1. a) Notation for message passing between HTM nodes. b) Graphical representation of the information processing within an intermediate node.

- Evidence coming from below is denoted e^-. In visual inference this is an image or video frame presented to level $\mathcal{L}_0$ of the network.
- Evidence from the top is denoted e^+ and can be viewed as contextual information. This can for instance be input from another sensor modality or the absolute knowledge given by the supervisor training the network.
- Feed-forward messages passed up the hierarchy are denoted λ and feedback messages flowing down are denoted π.
- Messages entering and leaving a node from below are denoted λ^- and π^- respectively, relative to that node. Following the same notation as for the evidence, messages entering and leaving a node from above are denoted λ^+ and π^+.

When the purpose of an HTM is that of a classifier, the feed-forward message of the output node is the posterior probability that the input e^- belongs to one of the problem classes. We denoted this posterior as $P(w_i|e^-)$, where w_i is one of n_w classes.

2.2   Internal Node Structure and Pre-training

HTM training is performed level by level, starting from the first intermediate level. The input level does not need any training, it just forwards the input. Intermediate levels training is unsupervised and the output level training is supervised. For a detailed description, including algorithm pseudocode, the reader should refer to [7].

For every intermediate node (Figure 1.b), a set C, of so called coincidence-patterns (or just coincidences) and a set, G, of coincidence groups, have to be learned. A coincidence, c_i, is a vector representing a prototypical activation pattern of the node's children. For a node in $\mathcal{L}_1$, with input nodes as children, this corresponds to an image patch of the same size as the node's receptive field. For nodes higher up in the hierarchy, with intermediate nodes as children, each element of a coincidence, $c_i[h]$, is the index of a coincidence group in child h. Coincidence groups, also called temporal groups, are clusters of coincidences likely to originate from simple variations of the same input pattern. Coincidences found in the same group can be spatially dissimilar but likely to be found close in time when a pattern is smoothly moved through the node's receptive field. By clustering coincidences in this way, exploiting the temporal smoothness of the input, invariant representations of the input space can be learned [9]. The assignment of coincidences to groups within each node is encoded in a probability matrix **PCG**; each element $PCG_{ji} = P(c_j|g_i)$ represents the likelihood that a group, g_i, is activated given a coincidence c_j. These probability values are the elements we will manipulate to incrementally train a network whose coincidences and groups have previously been learned and fixed.

The output node does not have groups but only coincidences. Instead of memorizing groups and group likelihoods it stores a probability matrix **PCW**, whose elements $PCW_{ji} = P(c_j|w_i)$ represents the likelihood of class w_i given the coincidence c_j. This is learned in a supervised fashion by counting how many times every coincidence is the most active one (the winner) in the context of each class. The output node also keeps a vector of class priors, $P(w_i)$, used to calculate the final class posterior.

2.3   Feed-Forward Message Passing

Inference in an HTM in conducted through feed-forward belief propagation (see [7]). When a node receives a set of messages from its m children, $\lambda^- = \{\lambda_1^-, \lambda_2^-, ..., \lambda_m^-\}$, a degree of certainty over each of the n_c coincidence in the node is computed. This quantity is represented by a vector y and can be seen as the activation of the node coincidences. The degree of certainty over coincidence i is

$$y[i] = \alpha \cdot p(e^-|c_i) = \begin{cases} e^{-(\|c_i - \lambda^-\|^2/\sigma^2)}, & \textit{if node level} = 1 \\ \prod_{j=1}^{m} \lambda_j^-[c_i[j]], & \textit{if node level} > 1 \end{cases} \tag{1}$$

where α is a normalization constant, and σ is a parameter controlling how quickly the activation level decays when λ^- deviates from c_i.

If the node is an intermediate node, it then computes its feed-forward message λ^+ which is a vector of length n_g and is proportional to $p(e^-|G)$, where G is the set of all coincidence groups in the node and n_g the cardinality of G. Each component of λ^+ is

$$\lambda^+[i] = \alpha \cdot p(e^-|g_i) = \sum_{j=1}^{n_c} PCG_{ji} \cdot y[j] \tag{2}$$

where n_c is the number of coincidences stored in the node.

The feed-forward message from the output node, the network output, is the posterior class probability and is computed in the following way:

$$\lambda^+[c] = P(w_c|e^-) = \alpha \cdot \sum_{j=1}^{n_c} PCW_{jc} \cdot P(w_c) \cdot y[j] \tag{3}$$

where α is a normalization constant such that $\sum_{c=1}^{n_w} \lambda^+[c] = 1$.

2.4   Feedback Message Passing

The top-down information flow is used to give contextual information about the observed evidence. Each intermediate node fuses top-down and bottom-up information to consolidate a posterior belief in its coincidence-patterns [9]. Given a message from the parent, $\boldsymbol{\pi}^+$, the top-down activation of each coincidence, $\mathbf{z}$, is

$$\mathbf{z}[i] = \alpha \cdot P(\mathbf{c}_i | e^+) = \sum_{k=1}^{n_g} PCG_{ik} \cdot \frac{\boldsymbol{\pi}^+[k]}{\boldsymbol{\lambda}^+[k]} \tag{4}$$

The belief in coincidence i is then given by:

$$\mathbf{Bel}[i] = \alpha \cdot P(\mathbf{c}_i | e^-, e^+) = \mathbf{y}[i] \cdot \mathbf{z}[i] \tag{5}$$

The message sent by an intermediate node (belonging to a level $\mathcal{L}_h, h > 1$) to its the children, $\boldsymbol{\pi}^-$, is computed using this belief distribution. The ith component of the message to a specific child node is

$$\boldsymbol{\pi}^-[i] = \sum_{j=1}^{n_c} I_{\mathbf{c}_j}\left(\mathbf{g}_i^{(child)}\right) \cdot \mathbf{Bel}[j] = \sum_{j=1}^{n_c}\sum_{k=1}^{n_g} I_{\mathbf{c}_j}\left(\mathbf{g}_i^{(child)}\right) \cdot \mathbf{y}[j] \cdot PCG_{jk} \cdot \frac{\boldsymbol{\pi}^+[k]}{\boldsymbol{\lambda}^+[k]} \tag{6}$$

where $I_{\mathbf{c}_j}(\mathbf{g}_i^{(child)})$ is the indicator function defined as

$$I_{\mathbf{c}_j}\left(\mathbf{g}_i^{(child)}\right) = \begin{cases} 1, & if\ group\ \mathbf{g}_i^{(child)}\ is\ part\ of\ \mathbf{c}_j \\ 0, & otherwise \end{cases} \tag{7}$$

The top-down message sent from the output node is computed in a similar way:

$$\boldsymbol{\pi}^-[i] = \sum_{c=1}^{n_w}\sum_{j=1}^{n_c} I_{\mathbf{c}_j}\left(\mathbf{g}_i^{(child)}\right) \cdot \mathbf{y}[j] \cdot PCW_{jc} \cdot P(w_c | e^+) \tag{8}$$

Equation 6 and 8 will be important when we, in the next section, show how to incrementally update the **PCG** and **PCW** matrices to produce better estimates of the class posterior given some evidence from above.

3   Htm Supervised Refinement

This section introduces a novel way to optimize an already trained HTM. The algorithm, called HSR (Htm Supervised Refinement) shares many features with traditional backpropagation used to train multilayer perceptrons and is inspired by weight fine-tuning methods applied to other deep belief architectures [3]. It exploits the belief propagation equations presented above to propagate an error message from the output node back through the network. This enables each node to locally update its internal probability matrix in a way that minimizes the difference between the estimated class posterior of the network and the posterior given from above, by a supervisor.

Our goal is to minimize the expected quadratic difference between the network output posterior given the evidence from below, e^-, and the posterior given the evidence from above, e^+. For this purpose we employ empirical risk minimization resulting in the following loss function:

$$L(e^-, e^+) = \frac{1}{2}\sum_{c=1}^{n_w}\left(P(w_c|e^+) - P(w_c|e^-)\right)^2 \tag{9}$$

where n_w is the number of classes, $P(w_c|e^+)$ is the class posterior given the evidence from above, and $P(w_c|e^-)$ is the posterior produced by the network using the input as evidence (i.e., inference). The loss function is also a function of all network parameters involved in the inference process. In most cases e^+ is a supervisor with absolute knowledge about the true class w_{c^*}, thus $P(w_{c^*}|e^+) = 1$.

To minimize the empirical risk we first find the direction in which to alter the node probability matrices to decrease the loss and then apply gradient descent.

3.1 Output Node Update

For the output node which does not memorize coincidence groups, we update probability values stored in the **PCW** matrix, through the gradient descent rule:

$$PCW'_{ks} = PCW_{ks} - \eta \frac{\partial L}{\partial PCW_{ks}} \qquad k = 1..n_c, \ s = 1..n_w \qquad (10)$$

where η is the learning rate. The negative gradient of the loss function is given by:

$$-\frac{\partial L}{\partial PCW_{ks}} = -\frac{1}{2} \sum_{c=1}^{n_w} \frac{\partial}{\partial PCW_{ks}} \left(P(w_c|e^+) - P(w_c|e^-)\right)^2 =$$

$$= \sum_{c=1}^{n_w} \left(P(w_c|e^+) - P(w_c|e^-)\right) \frac{\partial P(w_c|e^-)}{\partial PCW_{ks}}$$

which can be shown (see Appendix A of [10] for a derivation) to be equivalent to:

$$-\frac{\partial L}{\partial PCW_{ks}} = \mathbf{y}[k] \cdot Q(w_s) \qquad (11)$$

$$Q(w_s) = \frac{P(w_s)}{p(e^-)} \left(P(w_s|e^+) - P(w_s|e^-) - \sum_{i=1}^{n_w} P(w_i|e^-)(P(w_i|e^+) - P(w_i|e^-)) \right) \qquad (12)$$

where $p(e^-) = \sum_{i=1}^{n_w} \sum_{j=1}^{n_c} \mathbf{y}[j] \cdot PCW_{ji} \cdot P(w_i)$. We call $Q(w_s)$ the error message for class w_s given some top-down and bottom-up evidence.

3.2 Intermediate Nodes Update

For each intermediate node we update probability values in the **PCG** matrix, through the gradient descent rule:

$$PCG'_{pq} = PCG_{pq} - \eta \frac{\partial L}{\partial PCG_{pq}} \qquad p = 1..n_c, \ q = 1..n_g \qquad (13)$$

For intermediate nodes at level $\mathcal{L}_{n_{levels}-2}$ (i.e., the last but the output level) it can be shown (Appendix B of [10]) that:

$$-\frac{\partial L}{\partial PCG_{pq}} = \mathbf{y}[p] \cdot \frac{\pi_Q^+[q]}{\lambda^+[q]} \qquad (14)$$

where π_Q^+ is the child portion of the message π_Q^- sent from the output node to its children, but with $Q(w_s)$ replacing the posterior $P(w_s|e^+)$ (compare Eqs. 15 and 8):

$$\pi_{\bar{Q}}[q] = \sum_{c=1}^{n_w} \sum_{j=1}^{n_c} I_{c_j}\left(\mathbf{g}_q^{(child)}\right) \cdot \mathbf{y}[j] \cdot PCW_{jc} \cdot Q(w_c) \tag{15}$$

Finally, it can be shown that this generalizes to all levels of an HTM, and that all intermediate nodes can be updated using messages from their immediate parent. The derivation can be found in Appendix C of [10]. In particular, the error message from an intermediate node (belonging to a level $\mathcal{L}_h, h > 1$) to its child nodes is given by:

$$\pi_{\bar{Q}}[q] = -\sum_{t=1}^{n_c} \sum_{f=1}^{n_g} I_{c_j}\left(\mathbf{g}_q^{(child)}\right) \cdot PCG_{tf} \cdot \frac{\partial L}{\partial PCG_{tf}} = \sum_{t=1}^{n_c} \sum_{f=1}^{n_g} I_{c_t}\left(\mathbf{g}_q^{(child)}\right) \cdot PCG_{tf} \cdot \mathbf{y}[t] \cdot \frac{\pi_Q^+[f]}{\lambda^+[f]} \tag{16}$$

These results allow us to define an efficient and elegant way to adapt the probabilities in an already trained HTM using belief propagation equations.

3.3   HSR Pseudocode

A *batch* version of HSR algorithm is provided hereafter.

```
HSR( S )
{  for each training example in S
   {  Present the example to the network and do inference   (eqs. 1,2,3)
      Accumulate  ∂L/∂PCWₖₛ  values for the output node       (eqs. 11,12)
      Compute the error message πQ⁻                          (eq. 15)

      for each child of the output node:
            call BackPropagate(child, πQ⁺)       (see function below)
   }
   Update PCW by using accumulated  ∂L/∂PCWₖₛ               (eq. 10)
   Renormalize PCW such that for each class wₛ, ∑ₖ₌₁ⁿᶜ PCWₖₛ = 1
   for each intermediate node
   {  Update PCG by using accumulated  ∂L/∂PCGₚq            (eq. 13)
      Renormalize PCG such that for each group gq, ∑ₚ₌₁ⁿᶜ PCGₚq = 1
   }
}


function BackPropagate(node, πQ⁺)
{  Accumulate  ∂L/∂PCGₚq  values for the node                (eq. 14)
   if (node level > 1)
   {  Compute the error message πQ⁻                         (eq. 16)
         for each child of node:
            call BackPropagate(child, πQ⁺)
   }
}
```

By updating the probability matrices for every training example, instead of at the end of the presentation of a group of patterns, an online version of the algorithm is obtained. Both batch and online versions of HSR are investigated in the experimental section. In many cases it is preferable for the nodes in lower intermediate levels to share memory, so called node sharing [7]. This speeds up training and forces all the nodes of the level to respond in the same way when the same stimulus is presented at different places in the receptive field. For a level operating in node sharing, **PCG** update (eq. 13) must be performed only for the master node.

4   Experiments

To verify the efficacy of the HSR algorithm we performed a number of experiments on the SDIGIT dataset [7]. SDIGIT patterns (16×16 pixels, grayscale images) are generated by geometric transformations of prototypes called primary patterns. The possibility of randomly generating new patterns makes this dataset suitable for evaluating incremental learning algorithms. By varying the amount of scaling and rotation we can also control the problem difficulty.

With $S_{sdigit\ Train}\langle n, s_{xmin}, s_{xmax}, s_{ymin}, s_{ymax}, r_{max}\rangle$ we denote a set of n patterns, including, for each of the 10 digits, the primary pattern and further $(n/10) - 1$ patterns generated by simultaneous scaling and rotation of the primary pattern according to random triplets $\langle s_x, s_y, r\rangle$, $s_x \in [s_{xmin}, s_{xmax}]$, $s_y \in [s_{ymin}, s_{ymax}]$, $r \in [-r_{max}, r_{xmax}]$. The creation of a test set $S_{sdigit\ Test}\langle n, s_{xmin}, s_{xmax}, s_{ymin}, s_{ymax}, r_{max}\rangle$ starts by translating each of the 10 primary pattern at all positions that allow it to be fully contained in the 16×16 window thus obtaining m patterns; then, for each of the m patterns, $(n/10) - 1$ further patterns are generated by transforming the pattern according to random triplets $\langle s_x, s_y, r\rangle$; the total number of patterns in the test set is then $m \times n/10$. Examples of generated patterns are shown in Figure 2.

Fig. 2. Example of SDIGIT patterns. Ten patterns for every class are shown.

Table 1 (reprinted from [7]) summarizes HTM performance on the SDIGIT problem and compares it against other well know classification approaches. HTM accuracy is 71.37%, 87.56% and 94.61% with 50, 100 and 250 training patterns, respectively: our goal is to understand if accuracy can be improved by HSR incrementally training. To this purpose we follow the procedure described below:

```
Generate a pre-training dataset S₀ = SsdigitTrain < n,0.70,1.0,0.7,1.0,40° >
Pre-train a new HTM on S₀       (as in [7], leading to Table 1 results)
for each epoch Eᵢ,i = 1..nₑ
{  Generate a dataset Sᵢ = SsdigitTest < 1000,0.60,1.1,0.6,1.1,45° >   (6,200 patterns)
   Test HTM on Sᵢ
   for each iteration Iᵢ,i = 1..nᵢ
      call HSR(Sᵢ)
}
Test HTM on Sₙₑ
```

Table 1. HTM compared against other techniques on SDIGIT problem (reprinted from [7]). Three experiments are performed with an increasing number of training patterns: 50, 100 and 250. The test set is common across the experiments and include 6,200 patterns. NN, MLP and LeNet5 refer to Nearest Neighbor, Multi-Layer Perceptron and Convolutional Network, respectively. HTM refers to a four-level HTM (whose architecture is shown in Figure 1 of [10]).

SDIGIT - test set: $S_{sdigit\ Test}$(1000,0.60,1.10,0.60,1.10,45°)		(6,200 patterns, 10 classes)				
Training set	Approach	Accuracy (%)		Time (hh:mm:ss)		Size
		train	test	train	test	(MB)
$S_{sdigitTrain}$ <50,0.70,1.0,0.7,1.0,40°> 1788 translated patterns	NN	100	57.92	< 1 sec	00:00:04	3.50
	MLP	100	61.15	00:12:42	00:00:03	1.90
	LeNet5	100	67.28	00:07:13	00:00:11	0.39
	HTM	100	71.37	00:00:08	00:00:13	0.58
$S_{sdigitTrain}$ <100,0.70,1.0,0.7,1.0,40°> 3423 translated patterns	NN	100	73.63	< 1 sec	00:00:07	6.84
	MLP	100	75.37	00:34:22	00:00:03	1.90
	LeNet5	100	79.31	00:10:05	00:00:11	0.39
	HTM	100	87.56	00:00:25	00:00:23	1.00
$S_{sdigitTrain}$ <250,0.70,1.0,0.7,1.0,40°> 8705 translated patterns	NN	100	86.50	< 1 sec	00:00:20	17.0
	MLP	99.93	86.08	00:37:32	00:00:03	1.90
	LeNet5	100	89.17	00:14:37	00:00:11	0.39
	HTM	100	94.61	00:02:04	00:00:55	2.06

In our experimental procedure we first pre-train a new network using a dataset S_0 (with n patterns) and then for a number of epochs we generate new datasets S_i and apply HSR. At each epoch one can apply HSR for more iterations, to favor convergence. However, we experimentally found that a good trade-off between convergence time and overfitting can be achieved by performing just two HSR iterations for each epoch. The classification accuracy is calculated using the patterns generated for every epoch but before the network is updated using those patterns. In this way we emulate a situation where the network is trained on sequentially arriving patterns.

4.1   Training Configurations

We assessed the efficacy of the HSR algorithm for different configurations:

- *batch* vs *online* updating: see Section 3.3;
- *errors* vs *all* selection strategy: in *errors* selection strategy, supervised refinement is performed only for S_i patterns that were misclassified by the current HTM, while in *all* selection strategy is performed over all S_i patterns;
- learning rate η: see Equations 10 and 13. One striking find of our experiments is that the learning rate for the output node should be kept much lower than for the intermediate nodes. In the following we refer to the learning rate for output node as η_o and to the learning rate for intermediate nodes as η_i. We experimentally found that optimal learning rates (for SDIGIT problem) are $\eta_o = 0.00005$ and $\eta_i = 0.0030$.

Figure 3.a shows the accuracy achieved by HSR over 20 epochs of incremental learning, starting from an HTM pre-trained with $n = 50$ patterns. Accuracy at epoch 1 corresponds to the accuracy after pre-training, that is about 72%. A few epochs of HSR training are then sufficient to raise accuracy to 93-95%. The growth then slow down and, after 20 epochs, the network accuracy is in the range [97.0-98.5%] for the different configurations. It is worth remembering that the accuracy reported for each epoch is always measured on unseen data.

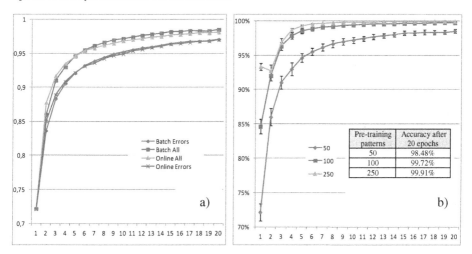

Fig. 3. a) HSR accuracy over 20 epochs for different configurations, starting with an HTM pre-trained with $n = 50$ patterns. Each point is the average of 20 runs. b) HSR accuracy over 20 epochs when using an HTM pre-trained with 50, 100 and 250 patterns. HSR configuration is *batch, all*. Here too HSR is applied two times per epoch. Each point is the average of 20 runs. 95% mean confidence intervals are plotted.

Training over all the patterns (with respect to training over misclassified patterns only) provides a small advantage (1-2 percentage). Online update seems to yield slightly better performance during the first few epochs, but then accuracy of online and batch update is almost equivalent. Table 2 compares computation time across different configurations.

Table 2. HSR computation times (averaged over 20 epochs). Time values refer to our C# (.net) implementation under Windows 7 on a Xeon CPU W3550 at 3.07 GHz.

Configuration	HSR time 6200 patterns - 1 iteration	HSR time 1 pattern - 1 iteration
Batch, All	19.27 sec	3.11 ms
Batch, Error	8,37 sec	1.35 ms
Online, All	22.75 sec	3.66 ms
Online, Error	8.27 sec	1.34 ms

Applying supervised refinement only to misclassified patterns significantly reduces computation time, while switching between *batch* and *online* configurations is not relevant for efficiency. So, considering that accuracy of the *errors* strategy is not far from the *all* strategy we recommend the *errors* configurations when an HTM has to be trained over a large dataset of patterns.

4.2   HTM Scalability

One drawback of the current HTM framework is scalability: in fact, the network complexity considerably increases with the number and dimensionality of training patterns. All the experiments reported in [7] clearly show that the number of coincidences and groups rapidly increases with the number of patterns in the training sequences. Table 3 shows the accuracy and the total number of coincidences and groups in a HTM pre-trained with an increasing number of patterns: as expected, accuracy increases with the training set size, but after 250 patterns the accuracy improvement slows down while the network memory (coincidences and group) continues to grow markedly, leading to bulky networks. Figure 3.b shows the accuracy improvement by HSR (*batch*, *all* configuration) for HTMs pre-trained over 50, 100 and 250 patterns. It is worth remembering that HSR does not alter the number of coincidences and groups in the pre-trained network, therefore the complexity after any number of epochs is the same for all the pre-trained HTMs (refer to Table 3). It is interesting to see that HTMs pre-trained with 100 and 250 patterns after about 10 epochs reach an accuracy close to 100%, and to note that even a simple network (pre-trained on 50 patterns) after 20 epochs of supervised refinement outperforms an HTM with more than 10 times its number of coincidences and groups (last row of Table 3).

Table 3. Statistics after pre-training. The first three rows are consistent with Table III of [7].

Number of pre-training patterns	Accuracy after pre-training	Coincidence and groups
50	71.37%	7193, 675
100	87.56%	13175, 1185
250	94.61%	29179, 2460
500	93.55%	53127, 4215
750	96.97%	73277, 5569
1000	97.44%	92366, 6864

5   Discussion and Conclusions

In this paper we propose a new algorithm for incrementally training HTM with sequentially arriving data. It is computationally efficient and easy to implement due to its close connection to the native belief propagation message passing of HTM.

The term $Q(w_s)$, the error message send from above to the output node (Eq. 12), is the information that is propagated back through the network and lies at the heart of the algorithm. Its interpretation is not obvious: the first part, $P(w_s|e^+) - P(w_s|e^-)$, the difference between the ground truth and network posterior, is easy to understand; while the second part, $-\sum_{i=1}^{n_w} P(w_i|e^-)(P(w_i|e^+) - P(w_i|e^-))$, is more mysterious. It is hard to give a good interpretation of this sum but from our understanding it arises due to the fact that we are dealing with probabilities. None of the parts can be ignored; tests have shown that they are both important to produce good results.

There are some parameters which need tuning to find the optimal setup. In the experiments presented in this paper two iterations per epoch were used, and the optimal learning rate was found therefor. With more iterations a lower learning rate would likely be optimal. The difference in suitable learning rate between the intermediate and the output level is also an important finding and can probably be explained by the fact that the **PCW** matrix of the output node has a much more direct influence on the network posterior. The output node memory is also trained supervised in the pre-training while the intermediate nodes are trained unsupervised, which might suggest that there is more room for fine tuning in the intermediate nodes. We ran some experiments where we only updated **PCW** in the output node: in this case a small performance gain of a few percent has been observed.

In general HSR has proven to work very well for the SDIGIT problem and the results give us reason to believe that this kind of supervised fine tuning can be extended to more difficult problems. Future work will focus on the following issues:

- applying HSR to other (more difficult) incremental learning problems;
- check whether, for a difficult problem based on a single training set, splitting the training set in two or more parts and using one part for initial pre-training and the rest for supervised refinement, can lead to better accuracy and efficiency;
- extending HSR in order to also finely tune (besides **PCW** and **PCG**) the structure of level 1 coincidences C without altering their number. In fact, while higher level coincidences are "discrete feature selectors" and therefore not applicable to continuous gradient descent optimization, level 1 coincidences are continuous features and their adaption could lead to further performance improvement.

References

1. George, D., Hawkins, J.: A Hierarchical Bayesian Model of Invariant Pattern Recognition in the Visual Cortex. In: IJCNN (2005)
2. Ranzato, M., et al.: Unsupervised Learning of Invariant Feature Hierarchies with Applications to Object Recognition. In: CVPR (2007)
3. Bengio, Y.: Learning Deep Architectures for AI. Foundations and Trends in Machine Learning 2(1) (2009)
4. Jarrett, K., Kavukcuoglu, K., Ranzato, M., LeCun, Y.: What is the Best Multi-Stage Architecture for Object Recognition? In: ICCV (2009)
5. George, D., Hawkins, J.: Towards a Mathematical Theory of Cortical Micro-circuits. PLoS Computational Biology 5(10) (2009)
6. Lee, T.S., Mumford, D.: Hierarchical Bayesian inference in the visual cortex. Journal of the Optical Society of America 20(7), 1434–1448 (2003)
7. Maltoni, D.: Pattern Recognition by Hierarchical Temporal Memory. DEIS TR (April 2011), http://bias.csr.unibo.it/maltoni/HTM_TR_v1.0.pdf
8. Pearl, J.: Probabilistic Reasoning in Intelligent Systems. Morgan-Kaufmann (1988)
9. George, D.: How the Brain Might Work: A Hierarchical and Temporal Model for Learning and Recognition. Ph.D. thesis, Stanford University (2008)
10. Maltoni, D., Rehn, E.M.: Incremental Learning by Message Passing in Hierarchical Temporal Memory. DEIS TR (May 2012), http://bias.csr.unibo.it/maltoni/HTM_HSR_TR_v1.0.pdf

Representative Prototype Sets
for Data Characterization and Classification

Ludwig Lausser*, Christoph Müssel*, and Hans A. Kestler**

Research Group Bioinformatics and Systems Biology
Institute of Neural Information Processing, Ulm University, 89069 Ulm, Germany
hans.kestler@uni-ulm.de

Abstract. Common classifier models are designed to achieve high accuracies, while often neglecting the question of interpretability. In particular, most classifiers do not allow for drawing conclusions on the structure and quality of the underlying training data. By keeping the classifier model simple, an intuitive interpretation of the model and the corresponding training data is possible. A lack of accuracy of such simple models can be compensated by accumulating the decisions of several classifiers. We propose an approach that is particularly suitable for high-dimensional data sets of low cardinality, such as data gained from high-throughput biomolecular experiments. Here, simple base classifiers are obtained by choosing one data point of each class as a prototype for nearest neighbour classification. By enumerating all such classifiers for a specific data set, one can obtain a systematic description of the data structure in terms of class coherence. We also investigate the performance of the classifiers in cross-validation experiments by applying stand-alone prototype classifiers as well as ensembles of selected prototype classifiers.

1 Introduction

The rapid development of molecular high-throughput technologies has driven the need for computational approaches to mine and analyze the resulting data. These data sets usually comprise only a small set of probes, while being extremely high-dimensional. An intuitive understanding of such data is usually impossible. Classifiers can provide decision support to life scientists when judging new probes. However, researchers are often interested in the basic characteristics that distinguish probes of different types. Feature selection techniques try to identify those features (e.g. genes) that are relevant for the classification.

Instead of selecting relevant features, we propose an approach which identifies probes that characterize certain classes well. This method is based on simple prototype set classifiers that comprise one sample from each class. Due to their simple structure and their data dependency, it is possible to enumerate all such classifiers for low-cardinality data sets, such as microarray data. This allows for a systematic characterization of the data set. For instance, the classification

---

* L. Lausser and C. Müssel contributed equally.
** Corresponding author.

N. Mana, F. Schwenker, and E. Trentin (Eds.): ANNPR 2012, LNAI 7477, pp. 36–47, 2012.
© Springer-Verlag Berlin Heidelberg 2012

performance of such simple base classifiers can give insight into the distribution and coherence of the classes. We describe several ways of extracting and visualizing information obtained from the universe of basic prototype set classifiers on a data set. Although single prototype set classifiers may be too simple to achieve a competitive classification accuracy, ensembles of such base classifiers that complement each other well can achieve a performance similar to state-of-the-art classifiers. We propose an ensemble method and evaluate it on several microarray data sets.

Prototype-based classification is used in many state-of-the-art classifiers, such as k-Nearest Neighbour (k-NN) classification [1], Learning Vector Quantization (LVQ) [2,3], Nearest Centroid classification, Nearest Medioid classification, or Nearest Shrunken Centroid classification [4]. With the exception of k-NN, these approaches generate prototypes based on the training set instead of directly drawing data points from the training set. Kuncheva and Bezdek [5] analyze whether prototypes should be selected or generated from the training set. They conclude that prototype selection should be preferred over prototype generation, as determining clusters in the data does not guarantee a good classification performance.

Our concept is related to approaches aiming at a reduction of the training set for the k-NN classifier. E.g., the Condensed k-NN approach tries to reduce the training subset in such a way that it maintains the performance of the full training set [6]. Many approaches make use of search heuristics, such as Genetic Algorithms (e.g. [7,8]). An overview of training subset selection for k-NN is given in [9] and [10].

Our approach differs from such neighbourhood condensation methods in several ways: Firstly, we focus on a very simple prototype representation using only a single prototype per class, while k-NN neighbourhood condensation can yield reduced training sets of arbitrary size. In this way, all possible classifiers can be examined, without the need of search heuristics to identify optimal sample subsets. Secondly, we consider these classifiers as base learners, i.e. they are not meant to achieve a high prediction accuracy on their own. Instead, they serve as data set descriptors and members of ensemble classifiers.

The structure of the paper is as follows: Section 2 describes the basic Representative Prototype Set (RPS) classifiers, their use for the characterization of data sets, and ensembles of RPS classifiers. In Section 3, the results of an application of the new methods to six well-known microarray data sets are presented. Section 4 discusses the results and concludes the paper.

2 Representative Prototype Set Classification

2.1 Basic Prototype Classifiers and Their Properties

We define a basic Representative Prototype Set (RPS) classifier as a set of prototypes, one for each class. The prototypes $\mathcal{P}$ are chosen from the training set. The labels of unseen data points are predicted according to the label of the nearest prototype in the set.

Let $\mathcal{T} = \bigcup_{i=1,\ldots,k} \mathcal{T}_i$ denote a labeled training set comprising k classes. A Representative Prototype Classifier consists of a set of prototypes

$$\mathcal{P} = \{p_i \in \mathcal{T}_i \mid i = 1, \ldots, k\},$$

where each $p_i = (\mathbf{x}_i, i)$ is a feature vector $\mathbf{x}_i \in \mathbf{R}^n$ labeled with class label i.

The classifier predicts an unseen data point $\mathbf{v}$ by choosing the prototype p_i with the smallest distance $d(\mathbf{v}, \mathbf{x_i})$, i.e.

$$RPS_\mathcal{P}(\mathbf{v}) = \text{argmin}_{i=1,\ldots,k}\ d(\mathbf{v}, \mathbf{x_i}).$$

In the following, we use the Euclidean distance.

The RPS classifier is a special case of general prototype classifiers PC which rely on a set of prototypes

$$\mathcal{Q} \subseteq \mathcal{T},$$

i.e, $\mathcal{Q}$ is not necessarily restricted to a single prototype per class.

The general prototype concept described above is data-dependent: A classifier c is called data-dependent if it can be determined entirely according to a relatively small set of training samples $\mathcal{T}' \subseteq \mathcal{T}$. That is, the training on both sets will result in the same classification model,

$$c_{\mathcal{T}'} = c_\mathcal{T} \tag{1}$$

The set of samples $\mathcal{T}'$ is called the *compression set* of the data-dependent classifier. For the above concept, the compression set is equal to $\mathcal{Q}$. If the prototype set corresponds to the complete training set ($\mathcal{Q} = \mathcal{T}$), the data-dependent prototype classifier corresponds to the well-known 1-Nearest Neighbour classifier [1].

Data-dependent classifiers allow for the specification of sample compression bounds [11]. These bounds can be used to give an upper limit of the true classification error probability $\mathcal{R}$ of a data-dependent classifier c. The main component of a sample compression bound is an empirical error rate R calculated on the remaining set of samples $\mathcal{T} \setminus \mathcal{Q}$.

Theorem 1 (e.g. [12]). *For a random sample $\mathcal{T}$ of iid examples drawn from an arbitrary, but fixed distribution $\mathcal{D}$ and for all $\delta \in (0, 1]$,*

$$\Pr_{\mathcal{T} \sim \mathcal{D}^{|\mathcal{T}|}}\left(\forall \mathcal{Q} \subseteq \mathcal{T} \text{ with } c = PC(x, \mathcal{Q}): \mathcal{R}_\mathcal{D}(c) \leq \overline{\text{Bin}}\left(\text{R}(c, \mathcal{T} \setminus \mathcal{Q}), \frac{\delta}{|\mathcal{T}|\binom{|\mathcal{T}|}{|\mathcal{T} \setminus \mathcal{Q}|}}\right)\right) \geq 1 - \delta$$

Here, $\overline{\text{Bin}}$ denotes the binomial tail inversion.

Proof. The sample compression bound given in Theorem 1 is a direct application of the sample compression bound in [12]. □

If there is a set of data-dependent prototype-based classifiers that all achieve the same empirical error rate, it follows from Theorem 1 that the classifier with the smallest compression set is most reliable. This means that in case of several classifiers with different numbers of prototypes, but equal performance, the extreme case of choosing the smallest possible reference set $\mathcal{P}$ as in RPS is the most favorable option.

2.2   Analyzing Data Set Characteristics

The total number of possible RPS classifiers for a training set $\mathcal{T}$ is $\prod_{i=1,\ldots,k} |\mathcal{T}_i|$. Consequently, it is often feasible to enumerate the complete set of RPS classifiers for a given data set, in particular regarding biomolecular data sets of low cardinality. This complete set can be used to characterize a data set according to the coherence of classes and representativeness of single training data points. For the following visualization and summarization approaches, we focus on two-class data sets.

For each RPS classifier c based on a prototype set $\mathcal{P}$, we can measure its empirical accuracy on the remaining training samples that are not included in the classifier:

$$A(c, \mathcal{T} \setminus \mathcal{P}) = 1 - R(c, \mathcal{T} \setminus \mathcal{P}) = \frac{|\{(\mathbf{v}, l) \in \mathcal{T} \setminus \mathcal{P} \mid c(\mathbf{v}) = l\}|}{|\mathcal{T} \setminus \mathcal{P}|}$$

For a two-class data set, this can be visualized in a heat map: the samples of the first class are plotted on the x axis, and the samples of the second class are plotted on the y axis. The greyscale color indicates the empirical accuracy of a combination, with a light color denoting a high accuracy and a dark color denoting a low accuracy. By applying complete-linkage hierarchical clustering, samples that exhibit a similar accuracy in combination with samples from the other class are grouped.

To get an impression of how well the data can be described by small sets of representative prototypes, we plot the distribution of the empirical accuracies A for all possible RPS classifiers c in form of a histogram. If many classifiers achieve high empirical accuracies, the histogram shows a right-skewed distribution, whereas data sets that are hard to separate by small prototype sets show a left-skewed distribution.

The empirical error rate of the classifiers is not the only performance measure that can be used to characterize a dataset. It is also of interest to know if all classifiers misclassify more or less the same set samples. A possible measure of this similarity of two classifiers c_a and c_b is Yule's Q statistic [13]:

$$Q_{i,j} = \frac{M^{11} M^{00} - M^{01} M^{10}}{M^{11} M^{00} + M^{01} M^{10}} \tag{2}$$

Here, M^{ij} denotes the number of co-occurences of correct predictions (1) or incorrect predictions (0) of c_a and c_b. E.g, M^{11} is the number of samples that are predicted correctly by both c_a and c_b, whereas M^{10} is the number of samples that are predicted correctly by c_a, but misclassified by c_b.

The range of Q is $[-1, 1]$. It is equal to 1 if both classifiers correctly predict exactly the same set of samples. It is equal to -1 if all samples correctly classified by c_a are misclassified by c_b and vice versa. If c_a and c_b are statistically independent, $\mathbb{E}[Q] = 0$.

Calculating Q for all pairs of possible RPS classifiers provides further information on a data set: If many pairs yield a Q close to 1, the data set is probably very coherent, and classes consist of a single cluster. By contrast, if many pairs achieve a Q close to -1, a prototypic description of the data set is not easily available, and classes probably consist of several disjoint clusters. A histogram of the distribution of Yule's Q on a data set can reveal such information.

2.3   Ensemble RPS Classification

A single RPS classifier can be seen as a simple base learner, but may not achieve good accuracies if any of the classes is distributed over two or more separate clusters due to the fact that each class is represented by only one prototype. To achieve a higher robustness and accuracy, several basic RPS classifiers can be combined to an ensemble classifier (denoted as eRPS). In the following, the ensemble set of m base learners is written as

$$\mathcal{E}_m = \{c_1, \ldots, c_m\}$$

To train an eRPS ensemble on a training data set $\mathcal{T}$, we determine all possible base RPS classifiers c_j and order them by their empirical accuracies $A(c_j, \mathcal{T})$. $\mathcal{E}_m$ then consists of the m RPS classifiers with the highest empirical accuracies.

An unseen sample $\mathbf{v}$ is classified according to a majority vote of the base learners, i.e.

$$eRPS_{\mathcal{E}_m}(\mathbf{v}) = \text{argmax}_{1,\ldots,k} \, | \, \{c_j \in \mathcal{E}_m \mid c_j(\mathbf{v}) = i\} \, |$$

3   Experiments

We applied our data set characterization as well as the ensemble classifier to several well-known microarray data sets:

The *Bittner data set* [14] contains expression profiles of 31 melanomas and 7 controls in 8067 features. The initial analysis of this data showed a stable cluster of 19 of the melanomas. In this analysis, the samples from this cluster (ML1) and the 19 remaining samples (melanomas and controls, ML2) were treated as distinct classes.

The *Golub data set* [15] contains data from a microarray experiment of acute Leukemia. The data set contains examples for two disease subtypes: ALL (acute lymphoblastic leukemia) and AML (acute myeloid leukemia). The 47 ALL and 25 AML examples consist of 3571 expression measurements. The probes were selected and normalized according to a procedure proposed by Dudoit et al. [16].

The *Notterman data set* [17] contains 18 paired samples of colon adeno-carcenomas (CA) and normal tissues (N). The expression profiles comprise 7457 features.

The *Pomeroy data set* [18] data set contains examples of two different kinds of embryonal tumors of the central nervous system, 25 classic medulloblastomas (CMD) and 9 desmoplastic medulloblastomas (DMD). The dataset contains 7129 unspecific genes.

The *Shipp data set*[19] consists of 77 samples of single B-cell lineage. 58 of these samples are classified as diffuse large B-cell lymphoma (DLBCL); the other

19 samples are follicular lymphoma (FL). The expression profiles of 7129 genes were collected using an Affymetrix HU6800 chip.

The *West data set* [20] comprises different breast cancer types. It contains 49 samples, which can be distinguished according to their estrogen receptor status (25 ER+ and 24 ER-). The expression profiles of 7129 features were measured using the HuGeneFL Chip.

For these six data sets, we include the heatmaps and accuracy histograms for single RPS classifiers as well as the histograms of Yule's Q statistic on pairs of RPS classifiers in Figures 1 and 2. The left columns of the figures show heatmaps of the accuracies of possible prototype sets in the data set. The greyscale color indicates the empirical accuracy of a combination, with a light color denoting a high accuracy and a dark color denoting a low accuracy. By applying a hierarchical clustering algorithm, samples that exhibit a similar accuracy in combination with samples from the other class are grouped. The columns in the middle show histograms of the same accuracies. The right columns depict histograms of the distribution of Yule's Q statistic for all pairs of prototype sets. This statistic measures how similarly these prototype sets predict the labels of samples.

To assess the classification performance of the RPS classifier, we conducted stratified cross-validation experiments, splitting the data into 10 random subsets, each of which was once used as a test set to measure the error, while the remaining samples were used for training. The cross-validation error was summed up over the 10 subsets, and the whole procedure was repeated 10 times. This yields a mean cross-validation error over the 10 runs.

Table 1 lists the mean cross-validation error percentage of the eRPS classifier, the k-Nearest Neighbour classifier, and the SVM with linear and RBF kernel for different configurations:

For the eRPS classifier, the number of prototype sets m in the ensemble was varied from 1 to 15. For k-NN, the number of neighbours k was varied, and for the linear SVM, different values of the cost parameter were applied. For the SVM with RBF kernel, we set the cost parameter to 100 (which appeared to yield the best results) and varied the γ parameter.

On the Bittner data set, eRPS clearly outperforms all other classifiers with only a single prototype set. The performance decreases when adding more sets to the majority vote, but remains clearly better than the results of the other classifiers. The accuracy histogram in Figure 1 can give a possible explanation for the good performance with few prototype sets: Most such base classifiers achieve a bad accuracy of 0.5 or less. However, there is a small number of prototype sets with an accuracy of more than 90%. As their performance is drastically better than nearly all other prototype sets, these good prototype sets are likely to be the top-ranked sets even in resampling settings, so that they are also included in the ensembles derived from the cross-validation subsets.This can also be seen in the heatmap: Many of the samples seem to be completely unsuitable as prototypes, as they yield a bad performance in any combination. Only a small set of configurations achieves a considerably higher performance.

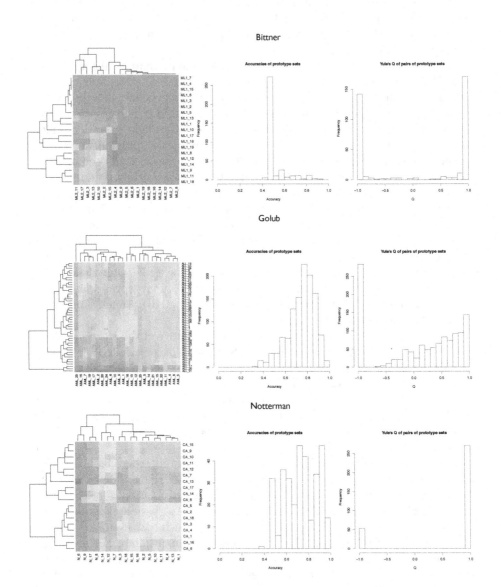

Fig. 1. Visualization of data set properties for the Bittner data set, the Golub data set and the Notterman data set

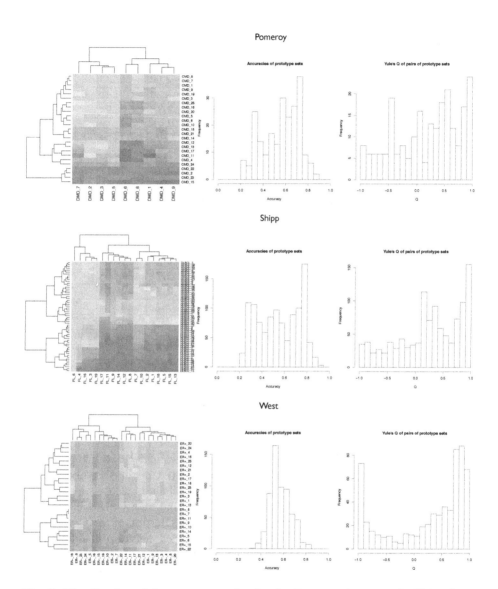

Fig. 2. Visualization of data set properties for the Pomeroy data set, the Shipp data set and the West data set

Table 1. Mean cross-validation prediction error percentage of Representative Prototype Set ensembles, k-Nearest Neighbour, and Support Vector Machines on six microarray data sets

	Bittner	Golub	Notterman	Pomeroy	Shipp	West
eRPS, $m = 1$	**7.63**	10.69	4.72	27.06	19.48	34.69
eRPS, $m = 3$	7.89	4.58	3.61	19.12	12.08	25.71
eRPS, $m = 5$	8.16	3.33	3.06	17.35	8.70	21.02
eRPS, $m = 7$	10.00	2.50	3.33	19.41	6.49	18.57
eRPS, $m = 9$	10.00	1.67	**2.78**	22.94	5.84	18.16
eRPS, $m = 11$	10.00	1.94	**2.78**	24.12	5.71	18.16
eRPS, $m = 13$	10.53	2.36	**2.78**	23.53	5.71	18.16
eRPS, $m = 15$	10.79	2.92	**2.78**	25.29	5.71	18.78
k-NN, $k = 1$	27.89	2.92	**2.78**	21.18	13.12	10.41
k-NN, $k = 3$	22.89	1.94	3.33	21.76	12.47	23.47
k-NN, $k = 5$	19.21	3.33	3.06	19.41	11.43	25.10
k-NN, $k = 7$	32.11	3.06	5.83	24.41	10.13	26.94
k-NN, $k = 9$	37.63	2.50	5.83	25.00	7.66	27.96
k-NN, k = 11	41.05	3.33	6.67	26.47	7.40	27.14
linear SVM, cost $= 0.001$	22.89	**1.39**	**2.78**	18.53	**3.77**	**8.98**
linear SVM, cost $= 0.01$	22.89	1.94	**2,78**	18.53	**3.77**	**8.98**
linear SVM, cost $= 0.1$	22.89	1.94	**2.78**	18.53	**3.77**	**8.98**
linear SVM, cost $= 1$	22.89	1.94	**2.78**	18.53	**3.77**	**8.98**
linear SVM, cost $= 10$	22.89	1.94	**2.78**	18.53	**3.77**	**8.98**
RBF SVM, cost $= 100$, $\gamma = 10^{-07}$	32.89	34.72	24.72	26.47	24.68	57.14
RBF SVM, cost $= 100$, $\gamma = 10^{-06}$	22.89	25.97	**2.78**	20.88	**3.77**	11.63
RBF SVM, cost $= 100$, $\gamma = 10^{-05}$	21.32	1.53	5.56	**16.47**	5.32	12.65
RBF SVM, cost $= 100$, $\gamma = 10^{-04}$	28.16	2.78	12.50	17.65	7.92	16.94
RBF SVM, cost $= 100$, $\gamma = 10^{-03}$	46.58	10.42	46.67	26.47	24.68	40.61

On the Golub data set, one configuration of the linear SVM achieves a slightly smaller error than the best eRPS classifiers, but both achieve an error of less than 2%. Here, more than one RPS classifier is required in the ensemble, with the best performance achieved for 9 base classifiers. This is also visible in the plots in Figure 1: The accuracy histogram shows that many RPS classifiers achieve good accuracies of 80% or more, while the distribution of the Q statistic indicates that there are many prototype sets that predict the samples differently. Hence, combining several good prototype sets that complement each other well can increase the performance. The heatmap indicates that the last three ALL samples are unsuitable as prototypes, while some other samples (e.g. AML_12, AML_13, AML_18, ALL_16, ALL_19, ALL_26) achieve good accuracies in almost any combination.

On the Notterman data set, the performance of all classifiers is similar, with the best error of 2.8% achieved by many configurations. The distribution of Yule's Q in Figure 1 reveals a possible reason: Most of the prototype classifiers seem to predict the same labels, while a small fraction of prototype sets behaves entirely different. Thus, ensembles comprising prototype classifiers of both categories will be able to improve the accuracy compared to single prototype classifiers, but the predictions of such ensembles will all be similar.

On the Pomeroy data set, the classification error is mostly similar for all three classifiers. The best accuracy is achieved by a configuration of the SVM with RBF kernel, followed by a configuration of 5 prototype sets of eRPS. The data set shows a broad variety of possible prototype sets whose classifications partly overlap, but are different for other samples (see the Q statistic histogram in Figure 2). Four samples in class CMD seem to be unsuitable prototypes (see heatmap), which indicates that this class is possibly incoherent.

On the Shipp data set, the SVM shows the best performance with an error of 3.8%. Some configurations of eRPS still achieve a very low error of 5.7%. Here, many basic prototype classifiers yield the same performance, but few achieve an accuracy of more than 90% (see Figure 2). Many of the classifiers predict similar labels. Interestingly, the heatmap shows that sample FL_6 yields a high accuracy in combination with any data point from the DLBCL class, which means that it is an excellent representative for his class.

On the West data set, eRPS is clearly outperformed by 1-NN and the SVM. Figure 2 shows that most prototype set classifiers achieve accuracies of 0.6 or less, such that this data set is probably unsuitable for prototype-based classification. At the same time, many of the base classifiers behave similarly, i.e. they misclassify the same samples. As a consequence, ensembles cannot benefit from a diverse set of base learners.

4   Discussion and Conclusion

The high dimensionality of current biomolecular data sets often makes an intuitive understanding impossible. Data mining approaches can provide decision support for such data. However, such models are usually not designed for easy interpretation.

We describe very simple base classifiers that predict the labels of unseen data points according to a set of single prototypes. Due to the simple nature and the data dependency of the Representative Prototype Set classifier, it is possible to describe data sets systematically by enumerating all possible classifiers. Furthermore, ensembles of such basic classifiers yield a prediction accuracy similar to state-of-the-art approaches. We have shown how the two key components of this paper, data set analysis and ensemble classification, complement each other well and can give insights into the data structure. In principle, the proposed methods work for any other type of concept class that allows for enumerating all classifiers.

As for any other classification approach, there is "no free lunch" [21]: the RPS approach is particularly suitable for certain data distributions, but inappropriate for others. The data set analysis also provides a way of judging the ensemble classifier's suitability for a specific type of data. For example, both the Bittner data set – where eRPS performs excellently – and the West data set – where eRPS is inferior to other approaches – show characteristic profiles in the data set analysis.

Future work will include different distance and correlation measures as well as other ways of aggregating the votes of the base classifiers in the ensembles. Furthermore, the selection of ensembles could be modified in such a way that the correct and incorrect predictions of the base classifiers on the training set complement each other.

Acknowledgement. This work is supported by the Graduate School of Mathematical Analysis of Evolution, Information and Complexity at the University of Ulm (CM, HAK) and by the German federal ministry of education and research (BMBF) within the framework of the program of medical genome research (PaCa-Net; project ID PKB-01GS08) and GerontoSys (Forschungskern SyStaR).

References

1. Fix, E., Hodges, J.: Discriminatory Analysis: Nonparametric Discrimination: Consistency Properties. Technical Report Project 21-49-004, Report Number 4, USAF School of Aviation Medicine, Randolf Field, Texas (1951)
2. Kohonen, T.: Learning vector quantization. Neural Networks 1, 303 (1988)
3. Kohonen, T.: Learning vector quantization. In: Arbib, M. (ed.) The Handbook of Brain Theory and Neural Networks, pp. 537–540. MIT Press, Cambridge (1995)
4. Tibshirani, R., Hastie, T., Narasimhan, B., Chu, G.: Diagnosis of multiple cancer types by shrunken centroids of gene expression. PNAS 99(10), 6567–6572 (2002)
5. Kuncheva, L., Bezdek, J.: Nearest prototype classification: Clustering, genetic algorithms or random search? IEEE Transactions on Systems, Man, and Cybernetics C28(1), 160–164 (1998)
6. Hart, P.E.: The condensed nearest neighbor rule. IEEE Transactions on Information Theory 14, 515–516 (1968)
7. Kuncheva, L.: Fitness functions in editing k-nn reference set by genetic algorithms. Pattern Recognition 30(6), 1041–1049 (1997)
8. Gil-Pita, R., Yao, X.: Using a Genetic Algorithm for Editing k-Nearest Neighbor Classifiers. In: Yin, H., Tino, P., Corchado, E., Byrne, W., Yao, X. (eds.) IDEAL 2007. LNCS, vol. 4881, pp. 1141–1150. Springer, Heidelberg (2007)
9. Brighton, H., Mellish, C.: Advances in instance selection for instance-based learning algorithms. Data Mining and Knowledge Discovery 6(2), 153–172 (2002)
10. Dasarathy, B.: Nearest neighbor (NN) norms: NN pattern classification techniques. IEEE Computer Society Press (1991)
11. Littlestone, N., Warmuth, M.: Relating data compression and learnability (1986) (unpublished manuscript)
12. Langford, J.: Tutorial on practical prediction theory for classification. Journal of Machine Learning Research 6, 273–306 (2005)
13. Yule, G.: On the association of attributes in statistics: With illustrations from the material of the childhood society. Philosophical Transactions of the Royal Society of London. Series A, Containing Papers of a Mathematical or Physical Character 194, 257–319 (1900)

14. Bittner, M., Meltzer, P., Chen, Y., Jiang, Y., Seftor, E., Hendrix, M., Radmacher, M., Simon, R., Yakhini, Z., Ben-Dor, A., Sampas, N., Dougherty, E., Wang, E., Marincola, F., Gooden, C., Lueders, J., Glatfelter, A., Pollock, P., Carpten, J., Gillanders, E., Leja, D., Dietrich, K., Beaudry, C., Berens, M., Alberts, D., Sondak, V.: Molecular classification of cutaneous malignant melanoma by gene expression profiling. Nature 406(6795), 536–540 (2000)
15. Golub, T., Slonim, D., Tamayo, P., Huard, C., Gaasenbeek, M., Mesirov, J., Coller, H., Loh, M., Downing, J., Caligiuri, M., Bloomfield, C., Lander, E.: Molecular Classification of Cancer: Class Discovery and Class Prediction by Gene Expression Monitoring. Science 286(5439), 531–537 (1999)
16. Dudoit, S., Fridlyand, J., Speed, T.: Comparison of discrimination methods for the classification of tumors using gene expression data. Journal of the American Statistical Association 97(457), 77–87 (2002)
17. Notterman, D., Alon, U., Sierk, A.J., Levine, A.: Transcriptional gene expression profiles of colorectal adenoma, adenocarcinoma, and normal tissue examined by oligonucleotide arrays. Cancer Research 61(7), 3124–3130 (2001)
18. Pomeroy, S., Tamayo, P., Gaasenbeek, M., Sturla, L., Angelo, M., McLaughlin, M., Kim, J., Goumnerova, L., Black, P., Lau, C., Allen, J., Zagzag, D., Olson, J., Curran, T., Wetmore, C., Biegel, J., Poggio, T., Mukherjee, S., Rifkin, R., Califano, A., Stolovitzky, G., Louis, D., Mesirov, J., Lander, E., Golub, T.: Prediction of central nervous system embryonal tumour outcome based on gene expression. Nature 415(6870), 436–442 (2002)
19. Shipp, M., Ross, K., Tamayo, P., Weng, A., Kutok, J., Aguiar, R., Gaasenbeek, M., Angelo, M., Reich, M., Pinkus, G., Ray, T., Koval, M., Last, K., Norton, A., Lister, T., Mesirov, J., Neuberg, D., Lander, E., Aster, J., Golub, T.: Diffuse large b-cell lymphoma outcome prediction by gene-expression profiling and supervised machine learning. Nature Medicine 8(1), 68–74 (2002)
20. West, M., Blanchette, C., Dressman, H., Huang, E., Ishida, S., Spang, R., Zuzan, H., Olson, J.J., Marks, J.R., Nevins, J.R.: Predicting the clinical status of human breast cancer by using gene expression profiles. PNAS 98(20), 11462–11467 (2001)
21. Wolpert, D.: The lack of a priori distinctions between learning algorithms. Neural Computation 8(7), 1341–1390 (1996)

Feature Selection by Block Addition
and Block Deletion

Takashi Nagatani and Shigeo Abe

Kobe University
Rokkodai, Nada, Kobe, Japan
abe@kobe-u.ac.jp
http://www2.kobe-u.ac.jp/~abe

Abstract. In our previous work, we have developed methods for selecting input variables for function approximation based on block addition and block deletion. In this paper, we extend these methods to feature selection. To avoid random tie breaking for a small sample size problem with a large number of features, we introduce the weighted sum of the recognition error rate and the average of margin errors as the feature selection and feature ranking criteria. In our methods, starting from the empty set of features, we add several features at a time until a stopping condition is satisfied. Then we search deletable features by block deletion. To further speedup feature selection, we use a linear programming support vector machine (LP SVM) as a preselector. By computer experiments using benchmark data sets we show that the addition of the average of margin errors is effective for small sample size problems with large numbers of features in realizing high generalization ability.

Keywords: Backward feature selection, feature ranking, forward feature selection, pattern classification, support vector machines.

1 Introduction

Feature selection aims at selecting the set of features, with the minimum number, that realizes a classifier with high generalization ability. In its original form, in backward selection, a feature is deleted sequentially from the full set of initial features or in forward selection, a feature is added sequentially to the initial empty set. If the recognition rate or more generally the generalization ability is used as the selection criterion, the selection method is called a wrapper method, and otherwise the filter method.

The wrapper method gives good generalization ability but its computational burden is high. In [1–3] the recognition rate for the cross-validation data set is used as a selection criterion. To speed up feature selection in such a situation, in [3], input variables for function approximation are added or deleted not one by one but in a block. In [4], a filter method and a wrapper method are combined to alleviate the computational burden of the wrapper method. In the filter stage, features with low class separability and high correlation with other features are

N. Mana, F. Schwenker, and E. Trentin (Eds.): ANNPR 2012, LNAI 7477, pp. 48–59, 2012.

eliminated. In the wrapper stage, a feature is added one by one by training the support vector machine (SVM) several times and selecting the feature with the maximum objective function value of the SVM.

Many filter methods have been developed and selection criteria based on mutual information are often used [5, 6]. As for SVM-based feature selection, the margin and the weights are widely used as a selection criterion. In [7], the SVM-RFE (SVM-Recursive Feature Elimination) is proposed in which backward feature selection is used with the minimum absolute weight in the separating hyperplane as a selection criterion.

As for sequential selection, a combination of forward selection with backward selection is proposed. In [8], sequential floating selection is proposed, in which after each sequential forward selection, backward selection is repeated so long as the selection criterion is satisfied.

With the introduction of SVMs, imbedded methods have been proposed, in which feature selection and training are done simultaneously. In [9], L0 norm, namely, the number of nonzero elements in the coefficient vector is included in the objective function. This term works to suppress irrelevant features. However, the generalization ability is usually inferior to regular SVMs. Therefore, in [10], like regular SVMs, the quadratic term is included in the objective function, in addition to the L0 or L1 norm term. This approach is extended to nonlinear cases.

To speed up wrapper methods, some feature selection methods use filter methods as a preselector such as LP SVMs with linear kernels [11, 3]. After training an LP SVM, input variables with small absolute values of weights are deleted.

In this paper, we extend the variable selection methods for function approximation [3] to pattern classification. The direct extension would replace, as a selection criterion, the approximation error with the recognition error rate. But because the recognition error rate is discrete, random tie breaking of feature ranking will occur for a large number of features and a small number of training data. To avoid this, we introduce the average of margin errors in addition to the recognition error rate.

The procedure for feature selection is almost the same as that in [3] excluding some alterations for improvement. We select features by block forward addition followed by block backward deletion. If the number of features is very large we use a filter method as a preselector.

Unlike feature ranking methods, the proposed feature selection method terminates when the selection criterion is no longer improved. Initially, we set the threshold value for the selection criterion using all the features. Then during feature selection we update the threshold value if the selection criterion better than the current threshold value is obtained. This guarantees to find the feature set with the selection criterion better than that of the initial set of features.

In block addition and block deletion, we simultaneously add or delete multiple features so long as the selection criterion is not worsened. Assuming that the number of features is large and also many irrelevant or redundant features are included, we fist do block addition then do block deletion because the former is more efficient than the latter is.

In Section 2, we discuss the selection criteria and the stopping conditions of feature selection. Then in Section 3 we discuss the proposed methods based on block addition and block deletion, and in Section 4, we show the simulation results using two-class benchmark data sets.

2   Selection Criteria and Stopping Conditions

In feature selection, we want to obtain a minimum feature set that realizes the generalization ability comparable to or better than that using the original feature set. From this point of view, the generalization ability estimated by cross-validation, namely, the recognition error rate for the validation data set that is hold out during cross-validation, is a good choice.

Let the decision function for a two class problem be trained so that

$$y_i \left(\mathbf{w}^\top \boldsymbol{\phi}(\mathbf{x}_i) + b \right) \geq 1 - \xi_i \qquad \text{for} \ \ i = 1, \dots, M, \tag{1}$$

where, $\mathbf{x}_i$ and y_i are the ith $(i = 1, \dots, M)$ training input and output, respectively, $\mathbf{w}$ is the coefficient vector of the separating hyperplane in the feature space, $\boldsymbol{\phi}(\mathbf{x})$ is the mapping function that maps $\mathbf{x}$ into the feature space, b is the bias term, and $\xi_i \, (\geq 0)$ is the slack variable associated with $\mathbf{x}_i$.

Then the recognition error rate E_C of the training data is given by

$$E_\mathrm{C} = \frac{1}{M} \sum_{i=1}^{M} e_i, \tag{2}$$

where

$$e_i = \begin{cases} 0 & \text{for} \ \ y_i \left(\mathbf{w}^\top \boldsymbol{\phi}(\mathbf{x}_i) + b \right) \geq 0, \\ 1 & \text{for} \ \ y_i \left(\mathbf{w}^\top \boldsymbol{\phi}(\mathbf{x}_i) + b \right) < 0. \end{cases} \tag{3}$$

The recognition error rate for the validation data set is calculated for the training data that are hold out in cross-validation.

Unlike the approximation error for function approximation, the recognition error rate is discrete. Therefore, if the number of features is large and the number of training data is small, the same recognition error rate will be obtained for different subsets of features. And random tie breaking during feature selection may not give a good selection result.

As a continuous criterion we consider using the average of margin errors:

$$E_\mathrm{M} = \frac{1}{M} \sum_{i=1}^{M} \xi_i, \tag{4}$$

where

$$\xi_i = \begin{cases} 0 & \text{for} \ \ y_i \left(\mathbf{w}^\top \boldsymbol{\phi}(\mathbf{x}_i) + b \right) \geq 1, \\ 1 - y_i \left(\mathbf{w}^\top \boldsymbol{\phi}(\mathbf{x}_i) + b \right) & \text{for} \ \ y_i \left(\mathbf{w}^\top \boldsymbol{\phi}(\mathbf{x}_i) + b \right) < 1. \end{cases} \tag{5}$$

Because we only want to use (4) to break ties, we consider the margin error-based criterion E_{Mc} as follows:

$$E_{\mathrm{Mc}} = E_{\mathrm{C}} + r\,E_{\mathrm{M}}, \tag{6}$$

where r is a positive parameter. Because the minimum positive change of E_{C} is $1/M$, assuming $E_{\mathrm{M}} \leq 1$ and $r \leq 1/M$, the ranking list of features by E_{Mc} and that by E_{C} are the same except for the feature subsets with the same E_{C}. Therefore we set $r = 1/M$.

To evaluate E_{Mc}, we consider the following three criteria for feature selection and feature ranking:

1. MM criterion: E_{Mc} for both feature selection and feature ranking;
2. CM criterion: E_{C} for feature selection and E_{Mc} for feature ranking;
3. CC criterion: E_{C} for both feature selection and feature ranking.

In the following, if there is no confusion, we simply say E instead of E_{Mc} or E_{C} and E is usually used for the selection criterion, and if it is used for the ranking criterion we denote it as *ranking E*.

At the start of feature selection, we set the threshold value for the selection criterion using all the features. Let the threshold be T. Then T is determined by $T = E^m$, where m is the number of initial features and E^m is the selection criterion evaluated by cross-validation. We update the threshold value when we obtain the selection criterion smaller than the threshold value as follows. Let the current selection criterion with j features be E^j. Then if

$$E^j < T, \tag{7}$$

we consider E^j as a new threshold value and set

$$T \leftarrow E^j \tag{8}$$

and continue feature selection. To obtain the smallest subset of features, we add features so long as $E < T$, namely we do not add features if $E = T$. And we delete features so long as $E \leq T$.

Now consider the difference among the three criteria for a small sample problem with a large number of features, where the problem is linearly separable. Suppose we obtain a subset of features with $E_{\mathrm{C}} = 0$ adding features using the CM or CC criterion. Because of the discussion above, we stop feature addition. But even if we obtain a subset of features with $E_{\mathrm{C}} = 0$ by feature deletion, we proceed feature deletion so long as $E_{\mathrm{C}} = 0$.

Now by the MM criterion, we add features so long as $E_{\mathrm{C}} = 0$ and E_{Mc} decreases. And we stop feature deletion, if E_{Mc} increases although $E_{\mathrm{C}} = 0$. Therefore, the MM criterion tends to select more features that the CM or CC criterion does.

To control that the MM criterion does not select much more features than the CM or CC criterion does, we consider two conditions to stop feature addition:

$$\Delta T < \varepsilon_{\mathrm{M}} \quad \text{or} \quad E_{\mathrm{C}} = 0, \tag{9}$$

where $\Delta T = T - E^j$ when (7) is satisfied and ε_{M} is a positive value.

3   Block Addition and Block Deletion

From the standpoint of quality of the selected feature set, backward selection, which deletes irrelevant or redundant features from the feature set, is more stable than forward selection, which selects features that are important only for the selected features. But if we need to select a small number of features from a large number of features, backward selection is slower than forward selection is. And this is usually the case for a large number of features.

To speed up feature selection in such a situation, in the following we discuss the method called feature selection by block addition and block deletion (BABD). In BABD, to speed up backward selection, we use forward selection as a pre-selector and afterwards, for the set of selected features we perform backward selection. To speedup forward and backward selection processes, we delete or add multiple features at a time and repeat addition or deletion until the stopping condition is satisfied.

3.1   Block Addition

First, we calculate the selection criterion E^m from the initial set of features $I^m = \{1, \ldots, m\}$ and set the threshold value of the stopping condition $T = E^m$. We start from the empty set of selected features. Assume that we have selected j features with the set of features I^j. Then we add the ith feature in set $I^m - I^j$ temporarily to I^j, calculate $E^j_{i_{\mathrm{add}}}$, where i_{add} indicates that the ith feature is added to the feature set, and calculate the selection criterion for the validation data set. Then we rank the features in $I^m - I^j$ in the ascending order of the ranking criteria. We call this ranking feature ranking V^j.

We add k $(k = 1, 2^1, \ldots, 2^A)$ features from the top of V^j to the feature set temporarily, where $2^A \le m$ and A is a user-defined parameter, which determines the number of added candidates. We compare E^{j+k} with the value of threshold T. If

$$E^{j+k} < T, \tag{10}$$

we update the threshold, add the features to the feature set permanently, and continue feature addition unless (9) is not satisfied for the MM criterion.

If (10) is not satisfied for $k = 1, 2^1, \ldots, 2^A$, we check if for some k the selection criterion is decreased by adding k features to I^j:

$$E^{j+k} < E^j. \tag{11}$$

Here we assume that $E^0 = \infty$. If it is satisfied let

$$k = \arg \min_{i=1,2^1,\ldots,2^A} E^{j+i} \tag{12}$$

and we add to I^j the first k features in the feature ranking permanently and continue feature addition.

If (10) and (11) are not satisfied, but $E^j \le T$, we stop feature addition. Otherwise we add to I^j the first feature in the feature ranking and continue feature addition to guarantee obtaining a feature set.

3.2   Block Deletion

Le the set of features obtained after block addition be I^j. Now by block deletion we delete redundant features from I^j. The reason for block deletion is as follows. In block addition, we evaluate feature ranking by temporarily adding one feature and we add multiple, high-ranked features. Thus, redundant features may be added by block addition.

We delete the ith feature in I^j temporarily from I^j and calculate $E^j_{i_{\mathrm{del}}}$, where $E^j_{i_{\mathrm{del}}}$ is the recognition error rate when we delete the ith feature from I^j. Then we consider the features that satisfy $E^j_{i_{\mathrm{del}}} \leq T$, as candidates of deletion and generate the set of features that are candidates for deletion by

$$S^j = \{i \mid E^j_{i_{\mathrm{del}}} \leq T, i \in I^j\}. \tag{13}$$

We generate V^j, ranking the candidates in the ascending order of *ranking* $E^j_{i_{\mathrm{del}}}$ and delete all the candidates from I^j temporarily. We compare $E^{j'}$ with the threshold T, where j' is the number of features after the deletion. If

$$E^{j'} \leq T, \tag{14}$$

block deletion has succeeded and we delete the candidate features permanently from I^j and update the threshold. If block deletion has failed, we backtrack and delete the features in the upper half of V^j. We iterate the procedure until block deletion succeeds.

We iterate the above procedure until no features are deleted.

3.3   Preselection by Filter Methods

If the number of features is very large, even block addition or block deletion may be inefficient. To overcome this problem we combine BABD with a filter method. We call this method BABD-FL. If the value of E for the subset of features obtained by the filter method is smaller than, or equal to, the threshold T, we update the threshold and delete features by block deletion from the set. If larger, we add features by block addition to the set of features obtained by preselection.

At first, we set the threshold of the stopping condition $T = E^m$ from the initial set of features I^m. By the filter method, we calculate the subset of features and set j_{FL} as the number of features after preselection. Then we compare the current selection criterion $E^{j_{\mathrm{FL}}}$ with the threshold T. If

$$E^{j_{\mathrm{FL}}} \leq T, \tag{15}$$

we update the threshold and search more deletable features by block deletion. If (15) is not satisfied, we add features to the current feature set by block addition until the selection criterion is below or equal to T. After block addition is finished we delete features by block deletion.

3.4   Algorithm of BABD-FL

In the following we show the algorithm of BABD-FL. If preselection is not used, we start from Step 4.

Preselection
Step 1 Calculate E^m for I^m. Set $T = E^m$.
Step 2 Calculate the subset of features by the filter method. Let the resulting subset of features obtained by preselection be $I^{j_{\mathrm{FL}}}$.
Step 3 Calculate $E^{j_{\mathrm{FL}}}$ for $I^{j_{\mathrm{FL}}}$ and set $j \leftarrow j_{\mathrm{FL}}$ and $E^j \leftarrow E^{j_{\mathrm{FL}}}$. If $E^{j_{\mathrm{FL}}} > T$, go to Step 5. Otherwise, go to Step 7.

Block Addition
Step 4 Calculate E^m for I^m. Set $T = E^m$, $j = 0$, and $E^0 = \infty$.
Step 5 Add the ith feature in $I^m - I^j$ temporarily to I^j, calculate *ranking* $E^j_{i_{\mathrm{add}}}$, and generate V^j. Set $k = 1$.
Step 6 Calculate E^{j+k} $(k = 1, 2^1, \ldots, 2^A)$. If (10) is satisfied, set $j \leftarrow j+k, T \leftarrow E^j$. And if (9) is satisfied for the MM criterion or selection is by BABD-FL go to Step 7; if not, go to Step 5. Otherwise, if (11) is satisfied, set $j \leftarrow j + k$ and go to Step 5. Otherwise, if both (10) and (11) are not satisfied but $E^j \leq T$, go to Step 7; otherwise set $j \leftarrow j + 1, T \leftarrow E^j$ and go to Step 5.

Block Deletion
Step 7 Delete temporarily the ith feature in I^j and calculate *ranking* $E^j_{i_{\mathrm{del}}}$.
Step 8 Calculate S^j. If S^j is empty, stop feature selection. If only one feature is included in S^j, set $I^{j-1} = I^j - S^j$, $j \leftarrow j - 1$ and go to Step 7. If S^j has more than two features, generate V^j and go to Step 9.
Step 9 Delete all the features in V^j from I^j: $I^{j'} = I^j - V^j$, where $j' = j - |V^j|$ and $|V^j|$ denotes the number of elements in V^j. Then, calculate $E^{j'}$ and if $E^{j'} > T$, go to Step 10. Otherwise, update j with j', $T \leftarrow E^{j'}$, and go to Step 7.
Step 10 Let V'^j include the upper half elements of V^j. Set $I^{j'} = I^j - \{V'^j\}$, where $\{V'^j\}$ is the set that includes all the features in V'^j and $j' = j - |\{V'^j\}|$. Then, if $E^{j'} \leq T$, delete features in V'^j and go to Step 7 updating j with j' and T with $E^{j'}$. Otherwise, update V^j with V'^j and iterate Step 10 until (14) is satisfied.

4   Performance Evaluation

In this section, we compare the three feature selection and feature ranking criteria for several two class problems. We also compare the proposed methods with the wrapper methods [2, 4] and the embedded method [10].

4.1   Evaluation Conditions

As a classifier we used a least squares SVM (LS SVM) whose primal problem is: minimize $1/2\mathbf{w}^\top \mathbf{w} + C/2 \sum_{i=1}^M \xi_i^2$ subject to $y_i (\mathbf{w}^\top \phi(\mathbf{x}_i) + b) = 1 - \xi_i$ for $i = 1, \ldots, M$, where C is the margin parameter. In training the LS SVM, we solved the set of linear equations that is derived by transforming the primal problem into the dual problem. As a kernel function, we used RBF kernels: $K(\mathbf{x}, \mathbf{x}') = \boldsymbol{\phi}^\top(\mathbf{x})\,\boldsymbol{\phi}(\mathbf{x}') = \exp(-\gamma\|\mathbf{x} - \mathbf{x}'\|^2)$, where γ is a positive parameter.

For BABD-FL, similar to the method in [3], we used the linear programming SVM (LP SVM) as a preselector and called the method BABD-LP. We selected the margin parameter in the LP SVM from $\{10, 10000\}$.

For BABD and BABD-LP, we set $A = 5$ and $\varepsilon_\mathrm{M} = 10^{-5}$.

We determine the γ and C values by fivefold cross-validation. We selected the γ value from $\{0.001, 0.01, 0.1, 0.5, 1, 5, 10, 15, 20, 50, 100\}$, and the C value from $\{1, 10, 50, 100, 500, 1000, 2000\}$. We trained the LS SVM for all pairs of parameter values and selected the values that realized the minimum value of the feature selection criterion for the validation data set.

For the MM criterion, the initial solution may be different from that for the CM and CC criteria. To make comparison easy, we set the same initial solution for the MM criterion as that for the CM and CC criteria.

To reduce computational cost of feature selection, first we determined the stopping threshold using all the features optimizing the γ and C values and then during feature selection, we fixed these parameter values to the determined values. This sped up feature selection without much deterioration of the generalization ability.

4.2   Comparison of Feature Selection Methods

We compared BABD-LP, BABD, and BD with the MM, CM, and CC criteria for two microarray problems, each of which consisted of 100 pairs of training and test data sets. Because these problems were linearly separable, we used linear kernels with $C = 1$. We calculated the average recognition rate for the test data set (validation data set), the average number of features selected by each method and its standard deviation, and measured the average computation time per data set using a personal computer with 3GHz CPU and 2GB memory.

Table 1 shows the results. In the "Data (Tr/Te/In)" column, we show the data set name followed by the numbers of training data, test data, and inputs in parentheses; and the recognition rates of the test data sets and, in parentheses, of validation data sets using all the features. In the "Method" column, LP, BA, and BD denote the BABD-LP, BABD, and BD, respectively. And BA* denotes that block addition was terminated when $E_\mathrm{C} = 0$. In the "Recognition rate" column, the best recognition rate among the three criteria is shown in boldface. In the "LP," "BA," and "BD" columns, we show the numbers of features selected by the LP SVM, BA, and BD, respectively, and the smallest average number of selected features in boldface. In the "Time (s)" column, we list the CPU time for feature selection per data set and the minimum time in bold face.

From the table, the MM criterion showed the best recognition rates for the six cases tested, although it required more features than the other two did, and thus, for BABD and BD, it required more computation time. If feature selection was terminated at $E_\mathrm{C} = 0$, feature selection was sped up but with the sacrifice

Table 1. Comparison of selection methods for microarray data sets

Data (Tr/Te/In)	Method	Recognition rate	LP	BA	BD	Time (s)
B. cancer	LP MM	**77.88**±10.44 (100±0)	**10.7**±1.0	0	8.4±1.5	**0.33**
(14/8/3226)	LP CM	70.50±13.41 (100±0)	**10.7**±1.0	0	2.0±0.4	**0.33**
73.88±11.47	LP CC	71.75±12.45 (100±0)	**10.7**±1.0	0	2.2±0.6	**0.33**
(76.50±7.09)	BA MM	**80.50**±11.36 (100±0)	—	46.0±11.3	40.5±11.9	6.98
	BA* MM	78.12±11.23 (100±0)	—	15.6±2.0	9.7±1.9	**0.42**
	BA CM	70.38±11.41 (100±0)	—	**2.4**±2.1	**1.7**±0.6	0.82
	BA CC	69.75±12.14 (99.93±0.71)	—	3.8±3.5	1.9±0.8	0.83
	BD MM	**78.88**±11.55 (100±0)	—	—	48.7±28.7	48.8
	BD CM	70.62±12.04 (89.14±13.87)	—	—	**1.6**±0.6	**48.0**
	BD CC	66.25±11.92 (85.93±12.22)	—	—	1.9±1.2	52.9
Leukemia	LP MM	**93.38**±3.98 (100±0)	**24.8**±2.3	0	16.1±2.5	**7.48**
(38/34/7129)	LP CM	87.97±5.97 (100±0)	**24.8**±2.3	0	**3.6**±0.9	7.56
94.44±4.70	LP CC	86.15±7.28 (100±0)	**24.8**±2.3	0	4.0±1.1	7.52
(92.45±3.32)	BA MM	**94.38**±3.88 (100±0)	—	57.9±12.6	47.9±12.2	68.0
	BA* MM	92.82±5.15 (100±0)	—	26.6±12.7	16.7±7.9	**13.8**
	BA CM	87.68±6.60 (99.79±0.71)	—	8.6±6.1	**3.4**±1.1	18.1
	BA CC	86.71±6.99 (99.21±1.26)	—	**8.3**±5.3	3.6±1.4	16.3
	BD MM	**94.68**±3.83 (99.97±0.26)	—	—	96.1±52.7	1862
	BD CM	87.82±6.99 (99.92±0.58)	—	—	**3.8**±0.9	**952**
	BD CC	81.85±7.45 (98.45±2.47)	—	—	6.2±2.9	1132

of the recognition rate (see BA* MM rows for the breast cancer and leukemia problems).

For BABD-LP, the average numbers of selected features in "BA" column are zero for the breast cancer and leukemia problems. This means that feature selection by LP SVM improved the selection criterion and no addition of features was necessary. In addition, the recognition rates for the validation data sets are all 100%. Therefore, preselection by LP SVM worked well.

By BA MM, the recognition rates for the validation data sets were all 100% while those for the CM and CC criteria were not. Even if $E_C = 0$ is reached, E_{Mc} may be positive for $E_C = 0$ and thus, by the MM criterion, feature addition may be continued after $E_C = 0$ so long as E_{Mc} is improved. Then at the block deletion stage, features will not be deleted if $E_C > 0$. But for the CM and CC criteria, this sort of thing is not satisfied.

By BD MM for the leukemia problem, the recognition rate for the validation data set is not 100%. This is because features were deleted before E_C reached 0.

The CM criterion showed better recognition rates than the CC criterion did except for LP CM for the breast cancer data set. Therefore, feature ranking by E_{Mc} worked better than by E_C for these data sets.

The generalization abilities of BABD and BD are comparable, but the feature selection time of BD is slower due to longer time for feature ranking.

Next, we compared BABD with other methods using the four data sets in [13]. As shown in [4], we randomly divided each data set into training and test data sets with the ratio of 80% and 20% and generated 20 pairs of training and

Table 2. Comparison of selection methods

Data (Tr/Te/In)	Method	Recognition rate	BA	BD
Bupa liver (276/69/6)	MM	**71.74**±5.79 (**73.13**±2.01)	**5.8**±0.5	5.7±0.6
72.68±6.14 (72.92±2.08)	CM, CC	71.67±5.71 (**73.13**±2.01)	**5.8**±0.6	**5.6**±0.7
	[4]	70.2	—	4.6
66.7±0.8	[10]	67.5±0.8	—	3.2
Ionosphere (281/70/34)	MM	**93.93**±2.59 (**97.10**±0.80)	25.1±4.6	**15.2**±5.0
94.21±1.89 (95.57±0.67)	CM	**93.93**±2.99 (96.83±0.64)	22.1±5.2	15.3±5.0
	CC	93.79±2.61 (96.74±0.72)	**21.6**±5.3	15.9±5.9
	[4]	92.0	—	10
92.9±0.2	[10]	92.3±0.3	—	6.6
Pima Indians (614/154/8)	MM	**75.39**±2.41 (**78.37**±0.55)	**6.5**±1.3	**6.0**±1.3
75.81±2.52 (77.81±0.82)	CM, CC	75.36±2.47 (78.33±0.56)	6.7±1.3	6.2±1.3
	[4]	74.5	—	4.2
76.6±0.2	[10]	73.0±0.2	—	1.4
WDBC (455/114/30)	MM	97.11±1.15 (**98.41**±0.33)	**23.4**±4.9	**16.6**±4.4
97.41±0.98 (98.09±0.34)	CM	**97.15**±1.07 (98.32±0.31)	24.1±6.0	21.4±7.0
	CC	97.11±1.30 (98.35±0.32)	24.4±5.9	22.4±6.6
	[4]	93.0	—	15
98.25±2.0	[2]	97.69±0.9	—	12

test data sets. But because the generated data sets, the number of data sets in some cases, and the classifiers used are different, exact numerical comparison is meaningless.

Table 2 shows the results. In [4], forward feature selection was done by the combination of filter and wrapper methods using the L1 SVM. In [10], the embedded method is used for feature selection. And in [2], sequential backward selection is used. The difference of the method with BD is that BD uses block deletion and threshold updating.

From the table, performance of BABD is better than or comparable to that of these methods.

Finally, using the two class data sets [12] with the numbers of features smaller than or equal to 60, we compared the feature selection criteria. We used BABD. The data sets consisted of 100 or 20 pairs of training and test data sets. Table 3 shows the results. The asterisk in the "Recognition rate" column means that the recognition rate is lower than that with all the features. The last to the third last rows of the table show the summary. For example, 6/1/5 in the MM row of the "Recognition rate" column means that the MM criterion performed best six times, the second best once, and the worst five times.

From the summary, it is interesting to note that the MM criterion showed the best recognition performance for the validation data sets but comparable with other two for the test data sets. But comparing the recognition rates, the difference among three criteria is very small for these data sets.

Although the feature selection methods guarantee the improvement of recognition rates for the validation data sets, it does not always lead to improvement in the recognition rates for the test data sets. Out of 12 problems, they deteriorated for 7 problems.

Table 3. Comparison of three selection criteria for two-class data sets

Data (Tr/Te/Im)	Method	Recognition rate	BA	BD
Cancer (200/77/9)	MM	73.14±4.35 (**77.90**±1.79)	5.8±2.6	4.4±1.8
	CM	73.22±4.45 (77.73±1.77)	**5.3**±2.7	**4.2**±2.0
73.05±4.61 (75.75±2.03)	CC	**73.35**±4.29 (77.74±1.77)	5.5±2.8	**4.2**±2.0
Diabetes (468/300/8)	MM	**76.08***±1.89 (78.64±1.08)	6.5±1.4	6.0±1.4
	CM	**76.08***±1.97 (**78.65**±1.08)	6.4±1.5	**5.9**±1.4
76.49±1.93 (78.00±1.18)	CC	**76.08***±1.92 (78.64±1.08)	6.4±1.5	**5.9**±1.5
Solar (666/400/9)	MM	**66.70**±2.07 (67.73±1.16)	5.9±2.5	4.7±1.9
	CM	**66.70**±2.06 (**67.74**±1.16)	5.3±3.0	4.5±2.5
66.29±1.95 (67.36±1.22)	CC	66.68±2.07 (67.73±1.14)	5.8±2.9	4.6±2.5
German (700/300/20)	MM	**75.46***±2.34 (**77.64**±0.96)	17.4±3.9	**13.6**±3.4
	CM	75.42*±2.21 (77.60±0.98)	**17.1**±4.1	13.9±3.9
75.93±1.89 (76.55±1.11)	CC	75.43*±2.00 (77.62±1.01)	18.9±2.5	13.9±3.5
Heart (170/100/13)	MM	82.05*±3.94 (85.81±2.04)	9.5±2.5	**8.2**±2.5
	CM	82.34*±3.78 (85.83±2.04)	**9.3**±2.8	8.3±2.7
82.48±3.61 (84.58±2.23)	CC	**82.45***±3.67 (**85.84**±2.03)	9.4±2.8	**8.2**±2.7
Image (1300/1010/18)	MM	97.86±0.34 (97.93±0.28)	**16.8**±0.6	**9.9**±3.0
	CM	**97.91**±0.38 (**97.97**±0.31)	17.2±0.7	10.9±2.1
97.69±0.56 (97.37±0.30)	CC	**97.91**±0.35 (97.94±0.30)	17.0±0.8	11.3±2.4
Ringnorm (400/7000/20)	MM	**97.55***±0.58 (**98.83**±0.51)	19.4±1.0	18.0±1.5
	CM	97.17*±0.79 (98.79±0.32)	**19.0**±1.3	**17.0**±2.0
98.11±0.27 (98.38±0.60)	CC	97.17*±0.79 (98.78±0.54)	**19.0**±1.3	17.0±2.1
Splice (1000/2175/60)	MM	92.50±1.12 (93.45±0.83)	9.4±4.4	8.6±4.3
	CM	92.46±1.06 (93.45±0.83)	9.3±4.5	8.5±4.3
89.05±0.81 (88.87±0.99)	CC	**92.56**±1.04 (**93.49**±0.81)	**9.2**±4.5	**8.3**±4.4
Thyroid (140/75/5)	MM	95.20*±2.45 (**97.40**±1.02)	**4.7**±0.5	**4.3**±0.8
	CM	95.25*±2.35 (97.38±1.01)	**4.7**±0.5	4.4±0.8
95.36±2.37 (97.14±1.03)	CC	**95.27***±2.37 (97.37±1.00)	**4.7**±0.5	4.4±0.8
Titanic (150/2051/3)	MM	**77.49**±0.67 (**79.41**±3.52)	**2.1**±0.9	**2.1**±0.9
	CM	**77.49**±0.67 (**79.41**±3.52)	2.4±0.8	2.4±0.8
77.43±0.77 (79.31±3.54)	CC	**77.49**±0.67 (**79.41**±3.52)	2.4±0.8	2.4±0.8
Twonorm (400/7000/20)	MM	**96.82***±0.73 (**98.29**±0.50)	18.7±1.7	18.3±1.6
	CM	96.48*±0.86 (98.24±0.58)	**18.3**±1.8	**17.5**±2.0
97.42±0.27 (98.00±0.58)	CC	96.52*±0.85 (98.28±0.53)	**18.3**±1.9	17.7±2.0
Waveform (400/4600/21)	MM	89.37*±1.17 (**92.30**±1.36)	19.0±1.6	**14.9**±2.8
	CM	**89.50***±1.13 (92.17±1.39)	18.8±1.7	15.5±2.8
90.09±0.58 (91.02±1.52)	CC	89.45*±1.16 (92.14±1.43)	**18.6**±2.0	15.6±2.7
Summary	MM	6/1/5 (7/3/2)	3/1/8	6/0/6
	CM	5/4/3 (4/5/3)	8/3/1	5/6/1
	CC	7/4/1 (3/6/3)	6/5/1	5/5/2

5   Conclusions

In this paper, we proposed a wrapper-based feature selection method by block addition and block deletion of features. Because feature selection and feature ranking by the recognition rate may cause random tie breaking especially for a large number of features and a small number of samples, we proposed using the weighted sum of the recognition error rate and the average of margin errors. The weight is determined so that the averages of margin errors do not change orders when there are no ties in the recognition rates. We select features first by adding several features at a time while the selection criterion is larger than the threshold value. Then, we delete several features at a time while the selection criteria is smaller than, or equal to, the threshold value. Initially, the threshold value is determined by using all the features. Then during feature selection, it is updated when the selection criterion lower than the threshold is obtained.

The computer experiments for two class data sets showed that the proposed selection criterion is better than the recognition error rate especially for the microarray data sets with a large number of features and a small number of samples.

References

1. Abe, S.: Modified backward feature selection by cross validation. In: Proc. ESANN 2005, pp. 163–168 (2005)
2. Maldonado, S., Weber, R.: A wrapper method for feature selection using support vector machines. Information Sciences 179(13), 2208–2217 (2009)
3. Nagatani, T., Ozawa, S., Abe, S.: Fast variable selection by block addition and block deletion. J. Intelligent Learning Systems & Applications 2(4), 200–211 (2010)
4. Liu, Y., Zheng, Y.F.: FS_SFS: A novel feature selection method for support vector machines. Pattern Recognition 39(7), 1333–1345 (2006)
5. Peng, H., Long, F., Dingam, C.: Feature selection based on mutual information: Criteria of max-dependency, max-relevance, and min-redundancy. IEEE Trans. Pattern Analysis and Machine Intelligence 27(8), 1226–1238 (2005)
6. Herrera, L.J., Pomares, H., Rojas, I., Verleysen, M., Guilén, A.: Effective Input Variable Selection for Function Approximation. In: Kollias, S.D., Stafylopatis, A., Duch, W., Oja, E. (eds.) ICANN 2006, Part I. LNCS, vol. 4131, pp. 41–50. Springer, Heidelberg (2006)
7. Guyon, I., Weston, J., Barnhill, S., Vapnik, V.: Gene selection for cancer classification using support vector machines. Machine Learning 46(1-3), 389–422 (2002)
8. Pudil, P., Novovičová, J., Kittler, J.: Floating search methods in feature selection. Pattern Recognition Letters 15(11), 1119–1125 (1994)
9. Bradley, P.S., Mangasarian, O.L.: Feature selection via concave minimization and support vector machines. In: Proc. ICML 1998, pp. 82–90 (1998)
10. Neumann, J., Schnörr, C., Steidl, G.: Combined SVM-based feature selection and classification. Machine Learning 61(1-3), 129–150 (2005)
11. Bi, J., Bennett, K.P., Embrechts, M., Breneman, C.M., Song, M.: Dimensionality reduction via sparse support vector machines. J. Machine Learning Research 3, 1229–1243 (2003)
12. IDA Benchmark Repository, http://www.fml.tuebingen.mpg.de/members/raetsch/benchmark
13. Asuncion, A., Newman, D.J.: UCI machine learning repository (2007), http://www.ics.uci.edu/~mlearn/MLRepository.html

Gradient Algorithms
for Exploration/Exploitation Trade-Offs:
Global and Local Variants

Michel Tokic[1,2] and Günther Palm[1]

[1] Institute of Neural Information Processing, University of Ulm, Germany
[2] Institute of Applied Research, University of Applied Sciences
Ravensburg-Weingarten, Germany

Abstract. Gradient-following algorithms are deployed for efficient adaptation of exploration parameters in temporal-difference learning with discrete action spaces. Global and local variants are evaluated in discrete and continuous state spaces. The global variant is memory efficient in terms of requiring exploratory data only for starting states. In contrast, the local variant requires exploratory data for each state of the state space, but produces exploratory behavior only in states with improvement potential. Our results suggest that gradient-based exploration can be efficiently used in combination with off- and on-policy algorithms such as Q-learning and Sarsa.

Keywords: reinforcement learning, exploration/exploitation.

1 Introduction

In reinforcement learning (RL), one of the most challenging tasks is balancing the amount of exploration and exploitation [1]. If the behavior of an agent is purely exploratory, the outcome of random actions prevents from maximizing short-term reward. In contrast, if an agent is purely exploitative, the selection of sub-optimal actions possibly prevents from maximizing long-term reward, because of underestimating the outcome of optimal actions. Conclusively, the optimal balance is somewhere in between, dependent on many parameters such as the learning rate, discounting factor, learning progress, and of course on the learning problem itself.

Many different approaches exist for tackling the trade-off between exploration and exploitation. Based on a single exploration parameter, some basic policies select random actions either equally distributed (ε-Greedy) or value sensitively (Softmax) [1], or by a combination of both [2], with the advantage of not requiring to memorize any exploratory data. In contrast, other approaches utilize counters for every state (exploration bonuses) direct the exploration process towards finding the optimal action-selection strategy in polynomial time under certain circumstances [3, 4]. Nevertheless, basic policies such as ε-Greedy and Softmax are known to be very effective having a proper exploration parameter configured, which has been successfully shown for example in board games with

N. Mana, F. Schwenker, and E. Trentin (Eds.): ANNPR 2012, LNAI 7477, pp. 60–71, 2012.
© Springer-Verlag Berlin Heidelberg 2012

huge discrete state spaces like Othello [5] or English Draughts [6]. In such state spaces, utility functions are hard to approximate and experiments for determining a proper exploration parameter can be time consuming. A non-convergent counter function is even harder to approximate than a convergent value function [7]. Interestingly, Daw et al. revealed in biologically-motivated studies on exploratory decisions in humans that there is [...] no evidence to justify the introduction of an extra parameter that allowed exploration to be directed towards uncertainty (softmax with an uncertainty bonus): at optimal fit, the bonus was negligible, making the model equivalent to the simpler softmax [8]. However, the search for an appropriate exploration parameter for such a policy remains as a challenging pattern-recognition task based on sensorimotor observations.

In the following, stochastic neurons are deployed for adapting exploration parameters by gradient-following algorithms. Adaptation of such parameters is with regard to the learning progress, instead of being tuned by hand in advance. We evaluate the presented approach in discrete and continuous state spaces using variants of the cliff-walking and mountain-car problems. The global variant of the presented algorithm was recently introduced in former research work [9]. Therefore, the contribution of this paper is an extended version in more detail also considering the local variant.

2   Methods

The learning problems considered in this paper can be described as Markov Decision Processes (MDP) that basically consist of a set of states, $\mathcal{S}$, and a set of possible actions within each state, $\mathcal{A}(s) \in \mathcal{A}, \forall s \in \mathcal{S}$ [1]. A stochastic transition function $\mathcal{P}(s, a, s')$ describes the (stochastic) behavior of the environment, i.e. the probability of reaching successor state s' after selecting action $a \in \mathcal{A}(s)$ in state s. The selection of an action is rewarded by a numerical signal from the environment, $r \in \mathfrak{R}$, used for evaluating the utility of the selected action. The goal of an agent is finding an optimal policy, $\pi^* : \mathcal{S} \to \mathcal{A}$, maximizing the cumulative reward. In the following, it is allowed for $\mathcal{S}$ to be continuous, but assumed that $\mathcal{A}$ is a finite set of actions. Action-selection decisions are taken at regular time steps, $t \in \{1, 2, \ldots, T\}$, until a maximum number of T actions is exceeded or a terminal state is reached.

2.1   Learning Algorithms

In temporal-difference learning, policies can be derived from utility functions representing so far learned knowledge [1]. An action-value function, $Q(s, a)$, denotes the cumulative and discounted reward for following policy π, when starting in state s and taking action a

$$Q(s, a) = E_\pi \left\{ \sum_{k=0}^{\infty} \gamma^k r_{t+k+1} | s_t = s, a_t = a \right\} , \qquad (1)$$

where $0 < \gamma \leq 1$ is a discounting factor used for weighting future rewards in $Q(s,a)$. Since $Q(s,a)$ depends on rewards received in the future, the cumulative reward is considered to be an expected value $E_\pi\{\cdot\}$.

Action-value function are learned from sensorimotor interactions of an agent with its environment. Two commonly used algorithms for learning $Q(s,a)$ are Sarsa for *on-policy* learning [7]

$$\Delta_{\text{Sarsa}} \leftarrow [r_{t+1} + \gamma Q(s_{t+1}, a_{t+1}) - Q(s_t, a_t)]$$
$$Q(s_t, a_t) \leftarrow Q(s_t, a_t) + \alpha \Delta_{\text{Sarsa}} , \tag{2}$$

and Q-learning for *off-policy* learning [10]

$$b^* \leftarrow \underset{b \in \mathcal{A}(s_{t+1})}{\arg\max} Q(s_{t+1}, b)$$
$$\Delta_{\text{Qlearning}} \leftarrow [r_{t+1} + \gamma Q(s_{t+1}, b^*) - Q(s_t, a_t)]$$
$$Q(s_t, a_t) \leftarrow Q(s_t, a_t) + \alpha \Delta_{\text{Qlearning}} , \tag{3}$$

where α denotes a stepsize parameter [11]. The technical difference between both algorithms is the inclusion of successor-state information used for evaluating action a_t in state s_t. Sarsa includes the discounted value of the actual selected action in the successor state, $Q(s_{t+1}, a_{t+1})$, for which reason the algorithm belongs to the family of on-policy algorithms. In contrast, Q-learning includes the discounted value of the estimated optimal action in the successor state, $Q(s_{t+1}, b^*)$, for which reason it is an off-policy algorithm. On-policy algorithms have the advantage of including into $Q(s,a)$ respective costs from stochastic action-selection policies, but have in turn no convergence guarantee, except when the policy has a greedy behavior [1].

2.2 Basic Exploration Policies

ε-**Greedy.** One basic policy for trading-off exploration/exploitation is selecting in state s an equally-distributed random action with probability $0 \leq \varepsilon \leq 1$, which is called an ε-Greedy policy. With probability $1 - \varepsilon$, a greedy action from the set of so far estimated optimal actions, $\mathcal{A}^*(s) = \arg\max_{a \in \mathcal{A}(s)} Q(s,a)$, is selected

$$\pi(\varepsilon, s, a) = \begin{cases} \frac{1-\varepsilon}{|\mathcal{A}^*(s)|} + \frac{\varepsilon}{|\mathcal{A}(s)|} & \text{if } a \in \mathcal{A}^*(s) \\ \frac{\varepsilon}{|\mathcal{A}(s)|} & \text{otherwise .} \end{cases} \tag{4}$$

Softmax. A disadvantage of ε-Greedy is that exploration actions are selected equally distributed among all possible actions, which may cause the income of high negative rewards from several bad actions, even if their true utility is correctly estimated. Therefore, an alternative is determining the selection probabilities according to a Boltzmann distribution (the Softmax policy), which also takes so far estimated utility into account

$$\pi(\tau, s, a) = \frac{e^{\frac{Q(s,a)}{\tau}}}{\sum_b e^{\frac{Q(s,b)}{\tau}}} . \tag{5}$$

Low settings of the exploration parameter τ (temperature) cause greediness, however high settings cause randomness.

MBE. A known problem of Softmax is that it *[...] has large problems of focusing on the best actions while still being able to sometimes deviate from them* [2]. This issue can be improved by combining Softmax with ε-Greedy into the *Max-Boltzmann Exploration* (MBE) rule [2], which selects exploration actions according to Softmax instead of being equally distributed

$$\pi(\varepsilon, \tau, s, a) = \begin{cases} \text{random action from } \mathcal{A}^*(s) \text{ with probability } 1 - \varepsilon \\ \text{Softmax action } \pi(\tau, s, a) \text{ with probability } \varepsilon \end{cases} \quad (6)$$

VDBE-Softmax. A drawback of the above policies is that an appropriate exploration parameter (ε or τ, or both for MBE) needs to be found for optimizing the cumulative reward. Such parameter varies dependently on learning progress, and typically an exploratory behavior is desired at the beginning of the learning process or in cases non-stationary environment responses are received. For this, some approaches make use of a decreasing exploration rate [12, 13], but which is known as to be inefficient for non-stationary environments. As a solution, we proposed the VDBE-Softmax policy in former research [14, 15], which adapts the exploration rate of MBE, state dependently based on fluctuations in the utility function

$$f(s, a, \phi) = 1 - e^{\frac{-|Q^n(s,a) - Q(s,a)|}{\phi}} = 1 - e^{\frac{-|\alpha \cdot \Delta|}{\phi}}$$
$$\varepsilon_{t+1}(s) = \delta \cdot f(s_t, a_t, \phi) + (1 - \delta) \cdot \varepsilon_t(s) , \quad (7)$$

where ϕ is a positive constant called *inverse sensitivity* and $\delta \in [0, 1)$ a parameter determining the effect of the selected action on the exploration rate. The second parameter (temperature) is set constantly to the value of $\tau = 1$, using normalized Q-values within the interval $[-1, 1]$.

3 Exploration-Parameter Control

Finding an appropriate exploration parameter by hand can be time consuming, and conclusively it is desired having algorithms adapting this parameter based on current operating conditions. The proposed solution for this problem is adapting the exploration parameter a_e of an action-selection policy, $\pi(a_e, \cdot, \cdot)$, towards improving the outcome of π based on some reasonable performance measure ρ. For maximizing ρ in the future, we deploy Williams' *"REINFORCE with multiparameter distributions"* algorithm using a stochastic neuron model [16], originally designed for reinforcement learning in continuous action spaces. The input of such neuron is a weighted parameter vector θ, from which the neuron determines the adaptable stochastic scalar as its output (the continuous-valued action). However, here we use discrete actions, thus the algorithm is applied for adapting the continuous-valued exploration parameter, i.e. REIN-FORCE Exploration-Parameter Control (REC). For example, if $\pi(a_e, \cdot, \cdot)$ is an ε-Greedy policy, the exploration parameter a_e refers to the exploration rate ε, i.e. $a_e \equiv \varepsilon$. In connectionist networks, such stochastic neurons can easily be integrated with compatibility to backpropagation [16]. The observed state s of the environment is typically the input to such network, which determines the parameter vector θ as its output to be processed by REC. However, in the following we use a tabular approximation of θ for the reason of measuring unbiased results.

At first, we show how a_e can be globally adapted with regard to the episodic cumulative reward, which is of interest for episodic learning problems consisting of a small set of starting states. Thereafter, a local variant is presented for determining a_e state dependently, aiming at producing exploratory behavior only in regions with improvement potential. However, the simple episodically version is an interesting variant because of being computational and memory efficient.

3.1   Global Episodic Control

In the following formulation, a single starting state s_s is assumed (for better readability), but which can easily be extended to multiple starting states as it will be discussed afterwards. The REC algorithm determines at each time step a continuous-valued action from a multiparameter distribution [16], representing the exploration parameter a_e. For this purpose, we use a normal (Gaussian) distribution with parameters μ (mean) and σ (standard deviation), which are given to the neuron as inputs.

At the beginning of each learning episode, the exploration parameter, being valid over the whole episode, is determined from the distribution (i.e. the activation function of the stochastic neuron), $a_e \sim \mathcal{N}(\mu, \sigma)$, whose density function g is given by

$$g(a_e, \mu, \sigma) = \frac{1}{\sigma\sqrt{2\pi}} e^{-(a_e - \mu)^2/2\sigma^2} \quad . \tag{8}$$

Let θ denote the vector of adaptable parameters consisting of

$$\theta = \begin{pmatrix} \mu \\ \sigma \end{pmatrix} \quad . \tag{9}$$

Our goal is adapting the components of θ at the end of episode i towards the gradient

$$\theta_{i+1} \approx \theta_i + \alpha\nabla_\theta\rho \quad . \tag{10}$$

For improving the future performance of $\pi(a_e, \cdot, \cdot)$, the policies outcome can be measured as the cumulative reward in episode i

$$\rho_i = E\{r_1 + r_2 + \cdots + r_T | \pi_i(a_e, \cdot, \cdot)\} \quad . \tag{11}$$

Next, the characteristic eligibility of each component of θ is estimated by

$$\frac{\partial \ln g(a_e, \mu, \sigma)}{\partial \mu} = \frac{a_e - \mu}{\sigma^2} \tag{12}$$

$$\frac{\partial \ln g(a_e, \mu, \sigma)}{\partial \sigma} = \frac{(a_e - \mu)^2 - \sigma^2}{\sigma^3} \quad , \tag{13}$$

and a reasonable algorithm for adapting μ and σ has the following form

$$\Delta\mu = \alpha_R(\rho - \bar{\rho})\frac{a_e - \mu}{\sigma^2} \tag{14}$$

$$\Delta\sigma = \alpha_R(\rho - \bar{\rho})\frac{(a_e - \mu)^2 - \sigma^2}{\sigma^3} \quad , \tag{15}$$

being applied at the end of each learning episode. The learning rate α_R has to be chosen appropriately, e.g. as a small positive constant, $\alpha_R = \alpha\sigma^2$, [16]. The baseline $\bar{\rho}$ is adapted by a simple reinforcement-comparison scheme

$$\bar{\rho} = \bar{\rho} + \alpha(\rho - \bar{\rho}) \quad . \tag{16}$$

Analytically, in Equation 14 the mean μ is shifted towards a_e in case of $\rho \geq \bar{\rho}$. On the contrary, μ is shifted towards the opposite direction if ρ is less than $\bar{\rho}$. Similarly, in Equation 15 the standard deviation σ is adapted in a way that the occurrence of a_e is increased if $\rho \geq \bar{\rho}$, and decreased otherwise (see proof in [16]). In simple words, the standard deviation controls exploration in the space of a_e.

A proper functioning of the algorithm depends on some requirements. The exploration parameter, mean, and standard deviation need to be bounded for obtaining reasonable performance (see Table 1). Furthermore, if the learning problem consists of more than one starting state, the parameters μ, σ and $\bar{\rho}$ must be associated to each occurring starting state, i.e. $\mu \to \mu(s), \sigma \to \sigma(s)$ and $\bar{\rho} \to \bar{\rho}(s)$, because way costs might affect ρ unevenly. However, if a learning problem consists of just one starting state, all utilized parameters can be considered as being global parameters.

Table 1. Parameter bounds for determining a_e

Policy	$\mu_{\min}$; $a_e^{\min}$	$\mu_{\max}$; $a_e^{\max}$	$\sigma_{\min}$	$\sigma_{\max}$
REC ε-Greedy: $\pi(\varepsilon, \cdot, \cdot)$	0.0	1.0	0.001	5.0
REC MBE: $\pi(\varepsilon, \cdot, \cdot)$	0.0	1.0	0.001	5.0
REC Softmax: $\pi(\tau, \cdot, \cdot)$	0.001	1000.0	0.1	5000.0
REC VDBE-Softmax: $\pi(\phi, \cdot, \cdot)$	0.001	1000.0	0.1	5000.0

3.2 Local Step-Wise Control

The results from the previous section can easily be extended for obtaining a local variant, aiming at producing exploratory behavior only in states with improvement potential. In general, all utilized parameters become local parameters associated to each state, i.e. $\mu \to \mu(s), \sigma \to \sigma(s)$ and $\bar{\rho} \to \bar{\rho}(s)$. The mean and standard deviation are readapted after evaluating action a performed in state s (by Q-learning or Sarsa). In prior to an action selection in state s, an exploration parameter a_e is determined based on $\mu(s)$ and $\sigma(s)$

$$a_e \sim \mathcal{N}(\mu(s), \sigma(s)) \ . \tag{17}$$

For evaluating the utility of a_e, the estimated utility of the actual taken action, $Q_{t+1}(s_t, a_t)$, is considered

$$\rho = Q_{t+1}(s_t, a_t) \ . \tag{18}$$

The following equations now readapt the distribution parameters from state s_t based on the policies outcome using its current parameter a_e

$$\Delta\mu(s_t) = \alpha_R(\rho - \bar{\rho}(s_t))\frac{a_e - \mu(s_t)}{\sigma(s_t)^2} \tag{19}$$

$$\Delta\sigma(s_t) = \alpha_R(\rho - \bar{\rho}(s_t))\frac{(a_e - \mu(s_t))^2 - \sigma(s_t)^2}{\sigma(s_t)^3} \ . \tag{20}$$

Finally, the baseline $\bar{\rho}(s)$ is readapted analogously to Equation 16

$$\bar{\rho}(s_t) = \bar{\rho}(s_t) + \alpha(\rho - \bar{\rho}(s_t)) \ . \tag{21}$$

4   Experiments

The presented policies are evaluated in two environments using *off-policy* Q-learning and *on-policy* Sarsa learning. First, a variation of the cliff-walking problem [1] is proposed as the *non-stationary cliff-walking problem* comprising a non-stationary environment. Second, a variation of the mountain-car problem is investigated comprising a continuous-valued state space approximated by a table, which causes partial observability of the actual coordinates. Investigated basic exploration policies are ε-Greedy, Softmax, MBE and VDBE-Softmax, using REC adaptation with parameter bounds according to Table 1. Since MBE requires two parameters to be set (ε and τ), we only adapt ε of this policy, while setting the temperature parameter constantly to the value of $\tau = 1$, and normalizing all Q-values in state s into the interval $[-1, 1]$. For the VDBE-Softmax policy, the inverse sensitivity parameter ϕ is adapted.

4.1   The Non-stationary Cliff-Walking Problem

The non-stationary cliff-walking problem is a modification of the cliff-walking problem presented by Sutton and Barto [1], which additionally comprises non-stationary responses of the environment. The goal for the agent is learning a path from start state S to goal state G_1, which is rewarded with the absolute

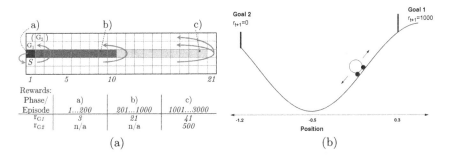

(a)   (b)

Fig. 1. The non-stationary cliff-walking problem (a) and the mountain-car problem with two goals (b)

costs of the shortest path minus 1 if successful (see Figure 1(a)). The reward for each action is defined as $r_{\text{step}} = -1$ (way costs). The environment also comprises unsafe *cliff states*, which lead to a high negative reward of $r_{\text{cliff}} = -100$, and also reset the agent back to the starting state S.

At the beginning of an experiment, learning takes place in phase (a) having one cliff state at left border. After 200 learning episodes, the grid world changes to phase (b), now comprising 10 cliff states. This change requires adapting the already learned behavior for circumventing the additional cliffs. After additional 800 episodes, the problem is tightened as shown in phase (c), where the number of cliffs is increased to 20. An alternative goal state $G2$ also appears, which is much higher rewarded with $r_{G2} = 500$ when entered.

Each episode begins in the starting state S, and terminates when either a goal state is entered or a maximum number of $T_{\max} = 200$ actions is exceeded. Throughout the experiment, the step-size parameter α is constantly set to the value of $\alpha = 0.2$. Since the learning problem is episodic, no discounting ($\gamma = 1$) is used. Finally, all action values are optimistically initialized with $Q_{t=0}(s,a) = 0$.

Results. Figure 2(a) shows the reward per episode averaged over 500 experiments. Averages of the mean and standard deviation for episodic policies are shown in Figure 2(b). It is observable that VDBE-Softmax maximizes the reward/episode in the episodic case. MBE shows best performance of the remaining three basic policies with the advantage of not requiring to memorize any further exploratory data such as utilized by VDBE-Softmax. The Sarsa algorithm shows better results for all four investigated policies when using episodic adaptation, but which is not the case for local adaptation. This discrepancy is for the reason that local adaptation tends to fast to become greedy, thus finding the second goal state in phase (c) more seldom. For comparison, all REC policies are better in episode 3000 compared to when using a pure greedy policy, which always converges to a reward per episode of 0, and which is only the optimal policy within the first 1000 episodes. In contrast, a pure random policy converges to -2750 respectively.

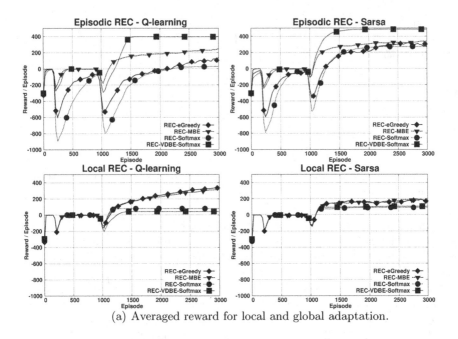

(a) Averaged reward for local and global adaptation.

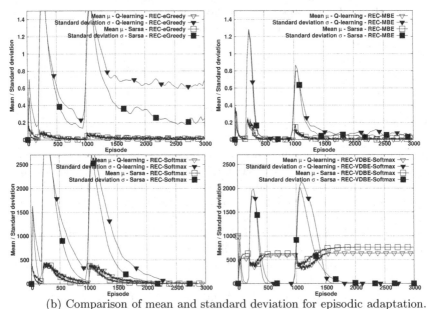

(b) Comparison of mean and standard deviation for episodic adaptation.

Fig. 2. The non-stationary cliff-walking problem: Averaged results (smoothed) for investigated REC policies using Q-learning and Sarsa. Note the dynamics of exploration for non-stationary environment responses.

4.2   The Mountain-Car Problem with Two Goals

In the mountain-car problem [17], the goal is driving an underpowered car up a mountain road, by initially standing in the valley between two mountains. The problem is that gravity is stronger than the car's engine, thus reaching the mountain top by full throttle only is not possible. Instead the car has to swing up for collecting enough inertia for overcoming gravity. In the here presented modification of the original learning problem, two goal states are utilized as depicted in Figure 1(b), which are rewarded differently upon arrival.

Once the car reaches one of the two goals, an episode terminates and the car is instantly set back to the middle of the valley. The idea of utilizing two differently valued goals is for the reason of measuring performance improvements achieved by various exploration policies, because a simple greedy policy (ε-Greedy with $\varepsilon = 0$) leads to optimal performance in the original description of the learning problem. The state variables are continuously valued consisting of the position of the car, $-1.2 \leq x \leq 0.3$, and its velocity, $-0.07 \leq \dot{x} \leq 0.07$. The dynamics of the environment are described by differential equations

$$x_{t+1} = bound\big[x_t + \dot{x}_{t+1}\big]$$
$$\dot{x}_{t+1} = bound\big[\dot{x}_t + 0.001a_t - 0.0025\cos(3x_t)\big] \ . \tag{22}$$

At each discrete time step, the agent can chose between one of seven actions, $a_t \in \{-1.0, -0.66, -0.33, 0, 0.33, 0.66, 1.0\}$, each rewarded by $r_{t+1} = -1$, except for reaching the right goal that is rewarded by $r_{t+1} = 1000$. An episode terminates when either one of the two goals has been arrived or when a maximum number of actions, $T_{\max} = 10000$, is exceeded. At the beginning of each episode the car is positioned in the valley at position $x = -0.5$ with initial velocity $\dot{x} = 0.0$. The state space is approximated by a 100×100 matrix, which causes the actual positions to be partially observable. Throughout the experiment, the step-size parameter α is constantly set to the value of $\alpha = 0.7$. Since the learning problem is episodic, no discounting ($\gamma = 1$) is used. Finally, all action values are optimistically initialized with $Q_{t=0}(s, a) = -200$.

Results. The results are averaged over 200 runs as shown in Figure 3. In general, best results are achieved for any policy using Q-learning in combination with local adaptation. Episodic adaptation of MBE outperforms ε-Greedy and Softmax. Furthermore, the Sarsa algorithm shows only to be advantageous in combination with ε-Greedy and Softmax using episodic adaptation, in contrast to MBE and VDBE-Softmax behaving more efficiently in combination with Q-learning. In the first phase of learning a degradation of performance is recognizable for episodic MBE and VDBE-Softmax. This is for the reason of learning at first a path to the left goal, and thereafter finding the right (better) goal. For comparison, a greedy policy converges to an average reward per episode of 114, in contrast to a pure random policy converging to -5440.

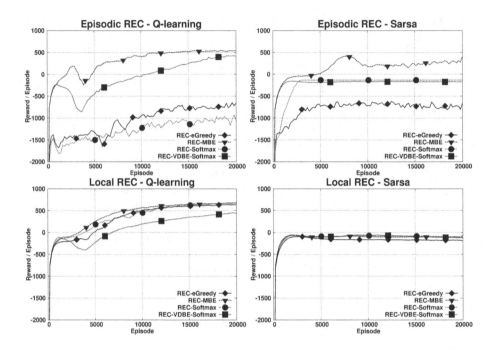

Fig. 3. The mountain-car problem with two goals: Averaged reward (smoothed) for investigated REC policies using Q-learning and Sarsa

5   Discussion and Conclusions

In this paper, new exploration policies are proposed for adapting the exploration parameter of basic exploration policies by gradient-following algorithms. Local and global variants have been evaluated in two different learning problems requiring to properly trade off exploration and exploitation. Results from the non-stationary cliff-walking problem show how the exploration parameter is readapted based on learning progress when non-stationary environment responses are received. For basic exploration policies in combination with REC, MBE shows to be most efficient in terms of performance. However, additional exploratory data might improve results as shown using episodic VDBE-Softmax in the non-stationary cliff-walking problem, but which is not always the case as it is observable in the mountain-car problem with two goals. When comparing local and global adaptation, the performance shows to be nearly the same for MBE, suggesting to use the memory and computational efficient global variant in episodic learning problems. Finally, the presented results show that gradient-following algorithms can be effectively used for balancing the exploration/exploitation dilemma inherent to reinforcement learning.

Acknowledgements. Michel Tokic received funding by the collaborative center for applied research *ZAFH-Servicerobotik*. The authors gratefully acknowledge the research grants of the federal state Baden-Württemberg and the European Union.

References

[1] Sutton, R.S., Barto, A.G.: Reinforcement Learning: An Introduction. MIT Press, Cambridge (1998)

[2] Wiering, M.: Explorations in Efficient Reinforcement Learning. PhD thesis, University of Amsterdam, Amsterdam (1999)

[3] Thrun, S.B.: Efficient exploration in reinforcement learning. Technical Report CMU-CS-92-102, Carnegie Mellon University, Pittsburgh, PA, USA (1992)

[4] Auer, P.: Using confidence bounds for exploitation-exploration trade-offs. The Journal of Machine Learning Research 3, 397–422 (2002)

[5] van Eck, N.J., van Wezel, M.: Application of reinforcement learning to the game of Othello. Computers and Operations Research 35, 1999–2017 (2008)

[6] Fauβer, S., Schwenker, F.: Learning a strategy with neural approximated temporal-difference methods in english draughts. In: Proceedings of the 20th International Conference on Pattern Recognition, ICPR 2010, pp. 2925–2928. IEEE Computer Society (2010)

[7] Rummery, G.A., Niranjan, M.: On-line Q-learning using connectionist systems. Technical Report CUED/F-INFENG/TR 166, Cambridge University (1994)

[8] Daw, N.D., O'Doherty, J.P., Dayan, P., Seymour, B., Dolan, R.J.: Cortical substrates for exploratory decisions in humans. Nature 441(7095), 876–879 (2006)

[9] Tokic, M., Palm, G.: Adaptive exploration using stochastic neurons. In: Proceedings of the 22nd International Conference on Artificial Neural Networks, Lausanne, Switzerland. Springer (to appear, 2012)

[10] Watkins, C.: Learning from Delayed Rewards. PhD thesis, University of Cambridge, England (1989)

[11] George, A.P., Powell, W.B.: Adaptive stepsizes for recursive estimation with applications in approximate dynamic programming. Machine Learning 65(1), 167–198 (2006)

[12] Grzes, M., Kudenko, D.: Online learning of shaping rewards in reinforcement learning. Neural Networks 23(4), 541–550 (2010)

[13] Nouri, A., Littman, M.L.: Multi-resolution exploration in continuous spaces. In: Koller, D., Schuurmans, D., Bengio, Y., Bottou, L. (eds.) Advances in Neural Information Processing Systems, vol. 21, pp. 1209–1216 (2009)

[14] Tokic, M., Palm, G.: Value-Difference Based Exploration: Adaptive Control between Epsilon-Greedy and Softmax. In: Bach, J., Edelkamp, S. (eds.) KI 2011. LNCS, vol. 7006, pp. 335–346. Springer, Heidelberg (2011)

[15] Tokic, M., Ertle, P., Palm, G., Söffker, D., Voos, H.: Robust Exploration/Exploitation Trade-Offs in Safety-Critical applications. In: Proceedings of the 8th International Symposium on Fault Detection, Supervision and Safety of Technical Processes, Mexico City, Mexico. IFAC (to appear, 2012)

[16] Williams, R.J.: Simple statistical Gradient-Following algorithms for connectionist reinforcement learning. Machine Learning 8, 229–256 (1992)

[17] Singh, S., Sutton, R.S.: Reinforcement learning with replacing eligibility traces. Machine Learning 22, 123–158 (1996)

Towards a Novel Probabilistic Graphical Model of Sequential Data: Fundamental Notions and a Solution to the Problem of Parameter Learning

Edmondo Trentin and Marco Bongini

Dipartimento di Ingegneria dell'Informazione
Università degli Studi di Siena, Siena, Italy
{trentin,bongini}@dii.unisi.it

Abstract. Probabilistic graphical modeling via Hybrid Random Fields (HRFs) was introduced recently, and shown to improve over Bayesian Networks (BNs) and Markov Random Fields (MRFs) in terms of computational efficiency and modeling capabilities (namely, HRFs subsume BNs and MRFs). As in traditional graphical models, HRFs express a joint distribution over a fixed collection of random variables. This paper introduces the major definitions of a proper dynamic extension of regular HRFs (including latent variables), aimed at modeling arbitrary-length sequences of sets of (time-dependent) random variables under Markov assumptions. Suitable maximum pseudo-likelihood algorithms for learning the parameters of the model from data are then developed. The resulting learning machine is expected to fit scenarios whose nature involves discovering the stochastic (in)dependencies amongst the random variables, and the corresponding variations over time.

Keywords: Probabilistic graphical model, Hidden Markov model, Hybrid Random Field, Sequence Classification.

1 Introduction

Probabilistic graphical models [9] have long been one of the hot topics in machine learning. The most popular instances are represented by directed graphical models, or Bayesian Networks (BNs) [10], and by undirected graphical models, namely Markov Random Fields (MRFs) [7]. Albeit intriguing and long-studied, BNs and MRFs present some drawbacks due to their very mathematical nature, and to certain limitations of the corresponding training algorithms [5]. In particular, the class $\mathcal{B}$ of (in)dependence structures that can be modeled via BNs, and the class $\mathcal{M}$ of (in)dependence structures that can be modeled via MRFs are such that $\mathcal{B} \cap \mathcal{M} \neq \varnothing$, $\mathcal{B} \not\subseteq \mathcal{M}$, and $\mathcal{M} \not\subseteq \mathcal{B}$. In other words, there are (in)dependence structures over any given set of random variables which can be modeled via BNs but not via MRFs, and vice-versa. Furthermore (and, possibly even more relevant in the computer science perspective), established learning and inference algorithms for BNs and MRFs present high computational complexity as the number of variables increases. Incidentally, no learning algorithm for

N. Mana, F. Schwenker, and E. Trentin (Eds.): ANNPR 2012, LNAI 7477, pp. 72–81, 2012.

MRFs has qualified as a standard reference so far. These are some of the reasons why real-world applications of BNs and MRFs have been somewhat limited to date. In [3], a new probabilistic graphical model was introduced with the aim of overcoming such drawbacks of traditional paradigms. The model, known as the Hybrid Random Field (HRF), was proven to subsume the modeling capabilities of both BNs and MRFs, meaning that the class $\mathcal{H}$ of (in)dependence structures that can be modeled via HRFs is such that $\mathcal{B} \subset \mathcal{H}$ and $\mathcal{M} \subset \mathcal{H}$ [2]. Moreover, HRFs come with clearly defined, efficient learning algorithms, that are empirically shown to yield models that (i) are at least as good as those obtained via BNs or MRFs, and that (ii) reduce the computational burden of learning (w.r.t. BNs and MRFs) to a dramatic extent [4].

HRFs, as well as traditional probabilistic graphical models (in particular, BNs and MRFs), assume a fixed set of random variables (whose joint probability distribution is actually modeled). Nonetheless, it is a fact that a number of real-world applications involve time-varying phenomena, presenting themselves in the form of (long, and often variable-length) sequences of time-dependent outcomes of a given set of random variables. Examples include a variety of problems in acoustic/speech processing (e.g., speech recognition, speaker identification, word-spotting, emotion recognition, etc.), video processing, bioinformatics (prediction of secondary and tertiary structure from observation of the primary structure of sequences of amino-acids, inference of functional properties from the primary structure, a variety of tasks in genomics and proteomics, etc.), natural language/document processing, etc. For this reason, interest in the development of dynamic extensions (i.e., suitable for sequential data) of the standard graphical models began to flourish in the scientific community. The most popular approach is represented by the dynamic BNs, proposed in [6]. It is noteworthy that standard hidden Markov models (HMMs) [11], which have long been applied to several amongst the aforementioned scenarios, can be shown to be a particular case of dynamic BNs [2]. Dynamic extensions of MRFs can be conceived, as well, although such extensions are not popular for the time being, and no training/inference algorithm has qualified as a standard so far. An alternative approach to the problem is the conditional random field (CRF) [8] which, under slightly different assumptions, found positive application to some of the tasks at hand, especially natural language and document processing.

The goal of the paper lies in attempting a first, formal definition of a proper dynamic extension of the standard HRF. As seen shortly, the definition involves latent random variables (in addition to the observable quantities) allowing for the modeling of (arbitrary length) sequences of sets of (time-dependent) random variables under Markov assumptions. Consequently, the resulting model is suitable for the recognition of sequential patterns as well, relying on the same classification strategy used in HRFs [2].

Suitable maximum pseudo-likelihood algorithms for learning (e.g., estimating the parameters of the resulting model) from a data sample are proposed, too. The model is expected to inherit the nice theoretical and computational properties of its static counterpart (the regular HRF), in particular as regards the positive comparison with respect to dynamic BNs and MRFs. In so doing, the practitioner is provided eventually with an alternative, effective tool for facing application tasks whose nature involves discovering the stochastic (in)dependencies amongst random variables and the corresponding variations over time. Although no experiments are presented herein, the companion paper [1] reports on comparative simulation results which corroborate the expectations empirically. The topic of structure learning in the novel graphical model is covered in [1], too.

The remains of the paper stretch out according to the following structure. Section 2 states the formal definition of the proposed model, by referencing to the original definition of HRF, and draws some preliminary conclusions on its modeling capabilities. Section 3 contextualizes properly the problem of learning from data, outlining the fundamental quantities involved in the calculations required in order to develop learning algorithms and the basic recursive scheme used in the following developments. Section 3.1 relies on these notions for coming up with a viable algorithmic solution to the problem of learning the parameters characterizing the different probabilistic distributions which constitute the model. Finally, Section 4 draws some preliminary conclusions.

2 Definitions

A broad-sense definition of HRFs is given in [2]. The reader is referred to the latter for all basic concepts of standard HRFs which are relevant to this paper. The modularity of an HRF is the property of factorizing the overall joint probability of its random variables in terms of a product of local probabilistic quantities defined at the level of the individual BNs embraced by the very definition of HRF. Noteworthily, modularity can be deduced as a property holding for HRFs defined accordingly [2]. A strict-sense, simpler definition can thence be devised, which roughly goes as follows: an HRF is a collection of Bayesian networks which possess the modular property (see, e.g., [3]). In this paper, for simplicity and coherence with the proposed algorithms, the strict-sense definition of HRF is assumed (without loss of generality). That said, we can give a definition of the novel dynamic probabilistic graphical model for sequences in the following terms. The model is referred to as the dynamic hybrid random field (DHRF).

Definition: a dynamic HRF $\mathcal{DH}$ is a tuple $\mathcal{DH} = (\mathbf{X}, S, \pi, \mathcal{F}, \mathbf{a}, \mathcal{H})$ where

1. $\mathbf{X}$ is a set of (observable) random variables $X_1, \ldots, X_n$. Outcomes of the random variables depend on time $t = 1, \ldots, T$, that is we will write $X_i(t)$ whenever we need to make the dependency explicit.
2. S is a set of Q latent radom variables, $S = \{S_1, \ldots, S_Q\}$. It is assumed that sequences of such latent variables are responsible for the generation of sequences of outcomes of the observable variables, and that the variables in S can be thought of as the states of a discrete-time Markov chain (*latent Markov assumption*). We write q_t to denote the state of the Markov chain at time t for $t = 0, \ldots, T$.

3. π is a probability distribution of the initial latent variables, i.e. $\pi = \{Pr(S_i \mid t = 0), S_i \in S\}$, where t is the discrete time index. For instance, if the Markov chain over S may equally-likely start with any latent variable, then π is uniform over S. Contrariwise, if a certain S_j can never occur at time $t = 0$, then $\pi(S_j) = 0$, etc.

4. $\mathcal{F} \subseteq S$ is the set of final states, i.e. the latent variables which can legitimately generate sets of outcomes of the observable variables at time T (namely, at the end of sequences).

5. $\mathbf{a}$ is a probability distribution that characterizes the (allowed) transitions between latent variables, that is $\mathbf{a}_{ij} = \{Pr(S_j$ at time $t \mid S_i$ at time $t - 1), S_i \in S, S_j \in S\}$ where the transition probabilities $\mathbf{a}_{ij}$ are assumed to be independent of time t. Note that the definition is meaningful due to the latent Markov assumption.

6. $\mathcal{H}$ is a set of HRFs over $\mathbf{X}$, $\mathcal{H} = \{\mathcal{H}_1, \ldots, \mathcal{H}_Q\}$, where $\mathcal{H}_q$ is uniquely associated with q-th latent variable S_q such that the joint emission probability $\mathbf{b}(\mathbf{X}) = P(X_1, \ldots, X_n \mid S_q)$ is modeled via HRF $\mathcal{H}_q$ over $\mathbf{X}$, independently of time t, and we assume that the probability distribution of $\mathbf{X}(t)$ is independent of the probability of $\mathbf{X}(t')$ (for all $t' \neq t$) given the latent variable (*emission Markov assumption*). In this definition, bearing in mind the definition of HRF, it turns out that $\mathcal{H}_q$ is a set of Bayesian networks $BN_{q,1}, \ldots, BN_{q,n}$ (with directed acyclic graphs $\mathcal{G}_{q,1}, \ldots, \mathcal{G}_{q,n}$) such that:

   (a) each $BN_{q,i}$ contains X_i plus a subset $\mathcal{R}_q(X_i)$ of $\mathbf{X} \setminus \{X_i\}$, namely the set of relatives of X_i in $BN_{q,i}$;
   (b) for each X_i, $P(X_i \mid \mathbf{X} \setminus \{X_i, q\}) = P(X_i \mid \mathcal{MB}_{q,i}(X_i))$, where $\mathcal{MB}_{q,i}(X_i)$ is the set containing the parents, the children, and the parents of the children of X_i in $\mathcal{G}_{q,i}$ (namely, the Markov blanket of X_i in $BN_{q,i}$).

   The Markov assumption holding in HRFs is referred to as the *observable Markov assumption* in the present framework.

Note that the overall DHRF can be thought of as a probabilistic graphical model over the set of random variables $S \cup \mathbf{X}$. Nonetheless, this definition allows for separate sets of BNs (i.e., different HRFs) for each latent variable, meaning that it does not extend regular HRFs to sequences by defining them as sets of dynamic Bayesian networks in a straightforward manner.

This definition is flexible and provides us with useful and efficient algorithmic tools rooted in traditional hidden Markov models (HMM). Before developing an algorithm for parameter learning in DHRFs, it is noteworthy to observe some fundamental properties regarding the modeling capabilities of DHRFs. Let $\mathcal{D}$ be the class of (in)dependence structures that can be represented via DHRFs. First of all, a DHRF with a single latent variable reduces implicitly to a regular HRF. Thence, $\mathcal{H} \subset \mathcal{D}$. Furthermore, since $\mathcal{B} \subset \mathcal{H}$ and $\mathcal{M} \subset \mathcal{H}$ (as we pointed out in Section 1), we have also $\mathcal{B} \subset \mathcal{D}$ and $\mathcal{M} \subset \mathcal{D}$. Given the fact that dynamic BNs (hence, including standard HMMs) are specialized instances of BNs, an immediate consequence of the reasoning is that DHRFs subsume the modeling capability of dynamic BNs (and, of HMMs). Following similar arguments, it is immediately seen that DHRFs subsume also any dynamic extensions (under Markov assumption) of Markov random fields.

3   Parameter Learning

Let $\mathcal{DH}$ be a DHRF and let $O = O_1, O_2, ..., O_T$ be a training sequence of outcomes of the observable variables, i.e. $O_t = (x_1, ..., x_n)$ for $t = 1, ..., T$. Training algorithms can be devised by exploiting a recursive scheme which is similar (to some extent) to the popular forward-backward procedures for HMMs [11].

The training criterion function is the (pseudo-)likelihood $P^*(O|\mathcal{DH})$, as in regular probabilistic graphical models (see [2] for a justification of why the pseudo-likelihood is used instead of the bare likelihood criterion). We define the (pseudo) forward terms

$$\alpha_t(i) = P(O_1, ..., O_t, q_t = S_i|\mathcal{DH}) \tag{1}$$

which can be recursively computed as

$$\alpha_t(i) = b_{i,t} \sum_j a_{ji}\alpha_{t-1}(j) \tag{2}$$

where $b_{i,t}$ denotes the emission probability of observation O_t given i-th latent variable. We say that the *forward step* of the following algorithms is the recursive computation of the α's according to Equation 2. Note that $P(O|\mathcal{DH}) = \sum_{i=1}^{Q} \alpha_T(i)$.

Also, the (pseudo) backward terms are defined as:

$$\beta_t(j) = P(O_{t+1}, ..., O_T|q_t = S_j, \mathcal{DH}) \tag{3}$$

which are recursively computed as

– initialization:

$$\beta_T(i) = \begin{cases} 1 \text{ if } S_i \in \mathcal{F} \\ 0 \text{ otherwise.} \end{cases} \tag{4}$$

– recursion:

$$\beta_t(j) = \sum_{i=1}^{Q} a_{ji}b_i(O_{t+1})\beta_{t+1}(i) \text{ for } t = T-1, T-2, ..., 1 \text{ and } 1 \leq j \leq Q \tag{5}$$

We refer to this recursive computation as the *backward step* of the following algorithms. We then define the quantity

$$\gamma_t(i) = P(q_t = S_i|O, \mathcal{DH}) \tag{6}$$

and we can write

$$\gamma_t(i) = \frac{P(q_t = S_i, O|\mathcal{DH})}{P(O|\mathcal{DH})} = \frac{P(O_1, ..., O_t, q_t = S_i, O_{t+1}, ..., O_T|\mathcal{DH})}{P(O|\mathcal{DH})} \tag{7}$$

$$= \frac{P(O_1, ..., O_t, q_t = S_i|\mathcal{DH})P(O_{t+1}, ..., O_T|O_1, ..., O_t, q_t = S_i, \mathcal{DH})}{P(O|\mathcal{DH})} \tag{8}$$

$$= \frac{P(O_1, ..., O_t, q_t = S_i|\mathcal{DH})P(O_{t+1}, ..., O_T|q_t = S_i, \mathcal{DH})}{P(O|\mathcal{DH})} \qquad (9)$$

$$= \frac{\alpha_t(i)\beta_t(i)}{\sum_{j=1}^{Q} P(q_t = S_j, O|\mathcal{DH})} \qquad (10)$$

$$= \frac{\alpha_t(i)\beta_t(i)}{\sum_{j=1}^{Q} \alpha_t(j)\beta_t(j)} \qquad (11)$$

where the emission Markov assumption was used to rewrite eq. 8 in the form of eq. 9. Put into words, $\gamma_t(i)$ can thus be calculated from the α's and β's during the backward step. Finally, we define:

$$\xi_t(i, j) = P(q_t = S_i, q_{t+1} = S_j \mid O, \mathcal{DH}) \qquad (12)$$

i.e.

$$\xi_t(i, j) = \frac{P(q_t = S_i, q_{t+1} = S_j, O \mid \mathcal{DH})}{P(O|\mathcal{DH})} \qquad (13)$$

$$- \frac{P(q_t = S_i, q_{t+1} = S_j, O|\mathcal{DH})}{\sum_{i=1}^{Q} \sum_{j=1}^{Q} P(q_t = S_i, q_{t+1} - S_j, O|\mathcal{DH})} \qquad (14)$$

$$= \frac{\alpha_t(i)a_{ij}b_j(O_{t+1})\beta_{t+1}(j)}{\sum_{i=1}^{Q} \sum_{j=1}^{Q} \alpha_t(i)a_{ij}b_j(O_{t+1})\beta_{t+1}(j)} \qquad (15)$$

which is computed during the backward step for $t = T, T - 1, ..., 1$, where $1 \leq i, j \leq Q$.

It is seen that the following properties hold true:

1. $\gamma_t(i) = \sum_{j=1}^{Q} \xi_t(i, j)$
2. $\sum_{t=1}^{T} \gamma_t(i)$ represents the expected number of instances of the latent variable S_i during the generation of the observed sequence O
3. $\sum_{t=1}^{T-1} \gamma_t(i)$ is the expected number of transitions starting from variable S_i (to any other variable)
4. $\sum_{t=1}^{T-1} \xi_t(i, j)$ is the expected number of transition that occur from S_i to S_j

Thus, an algorithm for re-estimating the probabilistic quantities involved in the definition of the DHRF $\mathcal{DH}$ and yielding a new (more accurate) DHRF $\widehat{\mathcal{DH}}$ involves the following formulas:

$$\widehat{\pi}_i = \gamma_1(i)$$
$$\widehat{a}'_{ij} = \frac{\sum_{t=1}^{T-1} \xi_t(i,j)}{\sum_{t=1}^{T-1} \gamma_t(i)} \text{ (from properties 3 and 4)}$$

which yield the re-estimation of initial and transition probabilities. The calculations accomplished so far underly the structure learning algorithm presented in the companion paper [1], as well. As regards re-estimation of the HRFs within the DHRF, an ad-hoc algorithm for parameter learning is covered in the next section.

3.1 Learning the Parameters of the HRFs within the DHRF

In this Section we propose a solution to the problem of learning the conditional probability tables (CPTs) [2] of the Bayesian networks modeling the local conditional distributions for each HRF $\mathcal{H}_1, \ldots, \mathcal{H}_Q$ associated with the latent variables. That is, for each latent variable $q = 1, \ldots, Q$ and for each observable variable X_i, the task is to learn the parameters of the conditional distribution $P(X_i|q, mb_{q,i}(X_i))$, for each state $mb_{q,i}$ of the variables in $\mathcal{MB}_{q,i}$, where $\mathcal{MB}_{q,i}$ is the Markov blanket [2] (in $\mathcal{H}_q$) for i-th observable variable and q-th latent variable. This requires that the structure of $\mathcal{H}_q$ has been previously fixed, i.e. that the directed acyclic graph (DAG) $\mathcal{G}_{q,i}$ associated in $\mathcal{H}_q$ with each variable X_i has been specified (the issue of learning an adaptive structure which is not fixed and pre-defined is covered in the companion paper [1]). In order to learn the parameters of each $BN_{q,i}$ from the training sequence $O = O_1, O_2, ..., O_T$, we use the technique described below. In order to denote the parents of node X_i in $\mathcal{H}_q$ we will use the notation $\mathcal{PA}_q(X_i)$, rather than using $\mathcal{PA}_{q,i}(X_i)$, since indexing the DAG only makes sense in the specific context of HRFs.

Let us assume that each observation O_j, $j = 1, \ldots, T$, is an n-dimensional vector $(x_{1_j}, \ldots, x_{n_j})$ of discrete values of the variables $X_1, \ldots, X_n$. The simplest case in parameter learning is the case of a variable X_i having no parents in the DAG within $\mathcal{H}_q$. In this case, we only need to estimate the absolute distribution $P(X_i|q)$. For each value x_{i_k} of X_i, our estimate reduces to the expected number of observations of x_{i_k} along the observed sequence O while being in the latent variable q, normalized by the expected number of presences in q, i.e.:

$$\widehat{P}(X_i = x_{i_k}|q) = \frac{\sum_{t, O_{t,k} = x_{i_k}} \gamma_t(q)}{\sum_t \gamma_t(q)} \tag{16}$$

where the sums over t are extended to $t = 1, \ldots, T$, $O_{t,k}$ denotes the k-th observed variable at time t, and the notation $\widehat{P}(X = x|q)$ refers to the new (improved) estimate of $P(X = x|q)$.

A more general case occurs in learning the conditional distribution of a node X_i in HRF $\mathcal{H}_q$ having parents $\mathcal{PA}_q(X_i)$. In this case, we need to estimate a distribution $P(X_i|pa_q(X_i))$ for each possible state $pa_q(X_i)$ of $\mathcal{PA}_q(X_i)$. For each value x_{i_k} of X_i, we will estimate these conditional probabilities from the training sequence O as the expected number of occurrences of outcome x_{i_k} for observable variable X_i jointly with $\mathcal{PA}_q(X_i) = pa_q(X_i)$ while being in latent variable q, normalized by the expected number of occurrences of the outcome $pa_q(X_i)$ of variables $\mathcal{PA}_q(X_i)$ while in q :

$$\widehat{P}(X_i = x_{i_k}|pa_q(X_i)) = \frac{\sum_{t, O_{t,k} = x_{i_k}, \mathcal{PA}_q(X_i) = pa_q(X_i)} \gamma_t(q)}{\sum_{t, \mathcal{PA}_q(X_i) = pa_q(X_i)} \gamma_t(q)} \tag{17}$$

The strategy proposed in Equations 16–17 has to be accomplished over the whole training set, i.e. not limited to an individual training sequences O (expectations need to be estimated over all training sequences), as in regular HMMs. Nonetheless, the technique suffers from the following problem. If a particular value x_{i_k} of X_i is never observed in the training set, or if it is never observed together with a particular configuration $pa_q(X_i)$ of $\mathcal{PA}_q(X_i)$, then our estimate of $P(X_i = x_{i_k}|q)$ (or of $P(X_i = x_{i_k}|q, pa_q(X_i))$) will be zero. This result is not acceptable in all cases where any event is possible, i.e. where every state of the network can be observed in principle. A solution for this difficulty appeals to the notion of an *equivalent state-occurrence expectation*, which we denote by $N^{(q)}$. It is the expected number of occurrences of latent variable S_q over a theoretical data sample which we assume to have been observed before the actual dataset. In other words, it is the expected number of occurrences of S_q over a *prior* sample. Within this prior sample, we assume to have observed any particular value x_{i_k} of X_i while in state q for a number of times equal to $p^{(q)} \cdot N^{(q)}$, where p stands for the prior probability that X_i has value x_{i_k} while in latent variable q.

Going back to Equations 16–17, we revise them as follows:

$$\widehat{P}(X_i = x_{i_k}|q) = \frac{\sum_{t, O_{t,k} = x_{i_k}} \gamma_t(q) + p_{i_k}^{(q)} N^{(q)}}{\sum_t \gamma_t(q) + N^{(q)}} \tag{18}$$

$$\widehat{P}(X_i = x_{i_k}|pa_q(X_i)) = \frac{\sum_{t, O_{t,k} = x_{i_k}, \mathcal{PA}_q(X_i) = pa_q(X_i)} \gamma_t(q) + p_{i_k}^{(q)} N_{pa_i}^{(q)}}{\sum_{t, \mathcal{PA}_q(X_i) = pa_q(X_i)} \gamma_t(q) + N_{pa_i}^{(q)}} \tag{19}$$

where $N_{pa_i}^{(q)}$ is a parameter related to $N^{(q)}$ in a way we will explain shortly. An important question concerning Equations 18–19 is what values we choose for the parameters $p_{i_k}^{(q)}$, $N^{(q)}$, and $N_{pa_i}^{(q)}$. In typical applications we assign uniform prior probabilities to the different values of each variable. Therefore, our choice for $p_{i_k}^{(q)}$ will be the following:

$$p_{i_k}^{(q)} = \frac{1}{|\mathcal{D}_i|} \tag{20}$$

where $\mathcal{D}_i$ is the domain of variable X_i. The value we assign to $N^{(q)}$ is instead:

$$N^{(q)} = \max_{1 \leq i \leq n} |\mathcal{D}_i| \tag{21}$$

An intuitive justification of Equation 21 appeals to two different aims. On the one hand, we want to keep the equivalent state-occurrence expectation as small as possible, so as to prevent prior probabilities from biasing learning too heavily. On the other hand, we want the equivalent state-occurrence expectation to be large enough to contain at least one occurrence for all values of each variable. Therefore, the choice made in Equation 21 seems to be an optimal trade-off. Given $N^{(q)}$, we define $N_{pa_i}^{(q)}$ as follows:

$$N_{pa_i}^{(q)} = \frac{N^{(q)}}{|\mathcal{D}_{\mathcal{PA}_{q,i}}|} \tag{22}$$

where $\mathcal{D}_{\mathcal{PA}_{q,i}}$ is the set of all possible states of $\mathcal{PA}_q(X_i)$. As a result of Equation 22, each value x_{i_k} of a non-root node X_i is expected to be observed $\frac{N^{(q)}}{|\mathcal{D}_i|}$ times within the prior (i.e., equivalent) sample, where $\frac{N^{(q)}}{|\mathcal{D}_i|} \geq 1$. To realize why $\frac{N^{(q)}}{|\mathcal{D}_i|} \geq 1$, we have to keep in mind that x_{i_k} is expected to be observed $p_{i_k}^{(q)} N_{pa_i}^{(q)}$ times for each possible state $pa_q(X_i)$ of $\mathcal{PA}_q(X_i)$. In other words, we have that $\frac{N^{(q)}}{|\mathcal{D}_i|} = p_{i_k}^{(q)} N_{pa_{q,i}}^{(q)} \cdot |\mathcal{D}_{\mathcal{PA}_{q,i}}|$.

4   Conclusions

As we said, probabilistic graphical models are a flexible, intriguing branch of machine learning. Traditional paradigms, being defined over fixed sets of random variables, are suitable for modeling joint distributions within "static" scenarios. A number of applications of the utmost interest, on the other hand, involve sequential data. This requires the capability of modeling time-varying stochastic (in)dependencies amongst random variables. The paper introduced a novel probabilistic graphical model for the modeling of sequences of random variables. Basically, its definition involves an underlying HMM structure combined with state-specific HRFs. This formulation is quite general under the Markov assumption, and subsumes dynamic Bayesian networks (including the traditional HMM itself) and any dynamic extensions of Markov random fields (provided that the time-dependencies satisfy Markov assumptions). An algorithm for learning the parameters of DHRFs was given, as well. The companion paper [1] focuses on the development of an algorithm for learning the structure of a DHRF, and presents empirical evidence (in the form of computer simulations) which corroborate the expectation that originally motivated the development of the present framework, that is: (i) having a dynamic graphical model which subsumes the capabilities of modeling (in)dependence structures offered by dynamic BNs and MRFs, whilst (ii) reducing their computational burden to a dramatic extent.

Acknowledgments. The authors would like to thank Antonino Freno for his invaluable support and the many stimulating discussions.

References

1. Bongini, M., Trentin, E.: Towards a Novel Probabilistic Graphical Model of Sequential Data: A Solution to the Problem of Structure Learning and an Empirical Evaluation. In: Mana, N., Schwenker, F., Trentin, E. (eds.) ANNPR 2012. LNCS (LNAI), vol. 7477, pp. 82–92. Springer, Heidelberg (2012)
2. Freno, A., Trentin, E.: Hybrid Random Fields: A Scalable Approach to Structure and Parameter Learning in Probabilistic Graphical Models. Springer (2011)
3. Freno, A., Trentin, E., Gori, M.: A Hybrid Random Field Model for Scalable Statistical Learning. Neural Networks 22, 603–613 (2009)
4. Freno, A., Trentin, E., Gori, M.: Scalable Pseudo-Likelihood Estimation in Hybrid Random Fields. In: Elder, J.F., Fogelman-Souli, F., Flach, P., Zaki, M. (eds.) Proceedings of the 15th ACM SIGKDD Conference on Knowledge Discovery and Data Mining (KDD 2009), pp. 319–327. ACM (2009)

5. Freno, A., Trentin, E., Gori, M.: Scalable Statistical Learning: A Modular Bayesian/Markov Network Approach. In: Proceedings of the International Joint Conference on Neural Networks (IJCNN 2009), pp. 890–897. IEEE (2009)
6. Ghahramani, Z.: Learning Dynamic Bayesian Networks. In: Giles, C.L., Gori, M. (eds.) IIASS-EMFCSC-School 1997. LNCS (LNAI), vol. 1387, pp. 168–197. Springer, Heidelberg (1998)
7. Kindermann, R., Laurie Snell, J.: Markov Random Fields and Their Applications. American Mathematical Society, Providence (1980)
8. Lafferty, J., McCallum, A., Pereira, F.: Conditional random fields: Probabilistic models for segmenting and labeling sequence data. In: Proc. 18th International Conf. on Machine Learning, pp. 282–289. Morgan Kaufmann, San Francisco (2001)
9. Lauritzen, S.L.: Graphical Models. Oxford University Press (1996)
10. Pearl, J.: Bayesian networks: A model of self-activated memory for evidential reasoning. In: Proceedings of the 7th Conference of the Cognitive Science Society, pp. 329–334. University of California, Irvine (1985)
11. Rabiner, L.R.: A tutorial on hidden markov models and selected applications in speech recognition. Proceedings of the IEEE 77(2), 257–286 (1989)

Towards a Novel Probabilistic Graphical Model of Sequential Data: A Solution to the Problem of Structure Learning and an Empirical Evaluation

Marco Bongini and Edmondo Trentin

Dipartimento di Ingegneria dell'Informazione,
Università degli Studi di Siena, Siena, Italy
{bongini,trentin}@dii.unisi.it

Abstract. This paper develops a maximum pseudo-likelihood algorithm for learning the structure of the dynamic extension of Hybrid Random Field introduced in the companion paper [5]. The technique turns out to be a viable method for capturing the statistical (in)dependencies among the random variables within a sequence of patterns. Complexity issues are tackled by means of adequate strategies from classic literature on probabilistic graphical models. A preliminary empirical evaluation is presented eventually.

Keywords: Probabilistic graphical model, Hidden Markov model, Hybrid Random Field, Sequence Classification.

1 Introduction

The notion of Dynamic Hybrid Random Field (DHRF) has been introduced in the companion paper [5], relying on the strict-sense definition of HRF given in [2] and summarized in [5]. The reader is referred to these bibliographical sources for all major concepts and the notation of standard HRFs and DHRFs. In short, an HRF is thence a model of the joint probability of a set of random variables that can be expressed in terms of a collection of Bayesian networks (BN) which possess the "modularity property".

This paper develops a maximum pseudo-likelihood algorithm for learning the structure of a Dynamic Hybrid Random Field from data. The technique is devised in Section 2. Once learning of the structure is completed, the algorithm for parameter learning presented in the companion paper [5] (coherently aimed at the maximization of the same criterion) may be applied.

The remains of the paper go as follows. Section 3 reports on a preliminary empirical evaluation of DHRFs involving synthetic sequences of patterns drawn from specific probability distributions. Therein, DHRFs are compared with Dynamic Bayesian Networks (DBN). Section 4 draws some preliminary conclusive remarks.

N. Mana, F. Schwenker, and E. Trentin (Eds.): ANNPR 2012, LNAI 7477, pp. 82–92, 2012.
© Springer-Verlag Berlin Heidelberg 2012

2   Structure Learning

Structure learning in q-th HRFs within the DHRF is the problem of learning, for each variable X_i, what other variables appear as nodes in the Bayesian network $BN_{q,i}$, and what edges are contained in the directed acyclic graph (DAG) $\mathcal{G}_{q,i}$. In other words, this means learning the structure of each Markov blanket $\mathcal{MB}_{q,i}(X_i)$ within q-th hybrid random field. While parameter learning assumes that the Markov blanket of each variable has previously been fixed, the aim of structure learning is to identify each Markov blanket and to determine its graphical structure.

We now present a heuristic structure learning algorithm for DHRFs, which we will call *dynamic Markov blanket merging* (DMBM). The aim of DMBM is to find an assignment of Markov blankets $\mathcal{MB}_{q,1}(X_1), \ldots, \mathcal{MB}_{q,n}(X_n)$ to the nodes $X_1, \ldots, X_n$ (within q-th HRF) that maximizes the model pseudo-likelihood given a dataset. The basic idea behind DMBM is to start from a certain assignment of neighbors to the variables of the model, learn the local Bayesian networks of the model, and then to iteratively refine the assignment so as to come up with Markov blankets that increase the model pseudo-likelihood with respect to the previous assignment. This iterative procedure stops when no further refinement of the Markov blankets assignment increases the value of the pseudo-likelihood. In other words, DMBM is nothing but a local search algorithm exploring a space of possible Markov blanket assignments to the observable variables for each state of the DHRF. An important problem to consider in this connection is the dimensionality of the search space. If we allow the search space to contain all possible Markov blanket assignments, the size of the space will be intractable: if n is the number of observable variables, for each variable there are 2^{n-1} possible Markov blankets, which means that the size of the search space will be $Q \cdot n \cdot 2^{n-1}$. Clearly, it is not possible to explore such a state space exhaustively. For this reason, DMBM is designed to explore only a small region of that space, as we are going to see.

In order to develop the algorithm, we specify three components: a way to produce the initial assignment, i.e. a model initialization strategy; a way to refine a given assignment so as to produce an alternative assignment, i.e. a search operator; a way to evaluate a given assignment, i.e. an evaluation function. Also, a general technique for learning the structure of Bayesian networks is required throughout the iterations of DMBM. The structure learning algorithm for BNs used in the paper, relying on the minimum description length principle, may be found in [2]. The next sections outline DMBM in detail.

2.1   Initial Segmentation

The first step of the algorithm relies on an initial segmentation of input sequences and on the creation of initial training sets for all HRFs $\mathcal{H}_1, \ldots, \mathcal{H}_Q$. These initial datasets are then used (see the next sections) for a heuristic initialization and refinement of the structures of the HRFs. Once the structures are learnt, the whole DHRF is used in order to obtain a new, improved segmentation of the original training sequences and the algorithm is iterated in an EM-fashion (e.g., relying on the Viterbi algorithm), yielding increasingly finer segmentations and structures. The initial segmentation can be obtained in several ways, e.g. relying on traditional machine learning approaches to sequence modeling. The simplest approach we consider herein involves a "linear" segmentation. Let us assume that a training sequence $O = O_1, \ldots, O_T$ is given, which has been generated by a sequence of latent variables $q_1, \ldots, q_L$ (note that $L << T$, in general). The linear segmentation splits O into L ordered subsequences having equal length $\ell = \lfloor T/L \rfloor$ (where $\lfloor . \rfloor$ denotes the usual floor function). State-specific datasets $\mathbf{D}_1, \ldots, \mathbf{D}_Q$ (initially empty) are built up from all the training sequences by merging them with the input observations that belong to the corresponding subsequences, i.e. $\mathbf{D}_1 = \mathbf{D}_1 \cup \{O_1, \ldots, O_\ell\}$, $\mathbf{D}_2 = \mathbf{D}_2 \cup \{O_{\ell+1}, \ldots, O_{2\ell}\}$, $\ldots$, $\mathbf{D}_Q = \mathbf{D}_Q \cup \{O_{(L-1)\ell+1}, \ldots, O_T\}$.

2.2 HRFs Initialization

The way DMBM produces an initial assignment (i.e., structure) for each $\mathcal{H}_q$ is by choosing an initial size k of the neighborhoods, and then by assigning as neighbors in $\mathcal{H}_q$ to each variable X_i those k variables that achieve the highest scores on the χ^2 dependence test over $\mathbf{D}_q$ with respect to X_i. The intuitive motivation for this choice is that a neighborhood containing variables that are more strongly correlated to X_i is more likely to capture the Markov blanket of X_i in $\mathcal{H}_q$ than a neighborhood containing variables that are only weakly correlated to X_i. Given the neighborhoods in $\mathcal{H}_q$, an assignment of Markov blankets to the respective variables is obtained by learning (both the structure and the parameters of) a Bayesian network $BN_{q,i}$ for each variable X_i, where $BN_{q,i}$ contains X_i together with its neighborhood $\mathcal{N}_q(X_i)$. As we said, in order to learn the structure of the local Bayesian networks, we use the technique described in [2].

2.3 Search Operator

Given current assignments of Markov blankets to the model variables in $\mathcal{H}_1, \ldots, \mathcal{H}_Q$, where the assignments are given by the Markov blankets $\mathcal{MB}_{q,1}(X_1), \ldots, \mathcal{MB}_{q,n}(X_n)$ specified by the networks $BN_{q,1}, \ldots, BN_{q,n}$ for $q = 1, \ldots, Q$, new assignments are obtained as follows. For each HRF $\mathcal{H}_q$ and for each variable X_i, we construct the set $\mathcal{U}_{q,i}$ as the union of $\mathcal{MB}_{q,i}$ with the Markov blankets of X_i in all graphs $\mathcal{G}_{q,j}$ such that X_i appears in $\mathcal{G}_{q,j}$ within the current assignment. Given the sets $\mathcal{U}_{q,1}, \ldots, \mathcal{U}_{q,n}$, we first check whether the cardinality of each $\mathcal{U}_{q,i}$ does not exceed a certain threshold k^*, and then we construct a new set of Bayesian networks $BN'_{q,1}, \ldots, BN'_{q,n}$ such that, for each q and for each i, if $|\mathcal{U}_{q,i}| \leq k^*$, then $BN'_{q,i}$ is the network learned by using the set $\mathcal{N}_q(X_i) = \mathcal{U}_{q,i}$ as neighborhood of X_i in $\mathcal{H}_q$, whereas, if $|\mathcal{U}_{q,i}| > k^*$, then $BN'_{q,i} = BN_{q,i}$. Given the networks $BN'_{q,1}, \ldots, BN'_{q,n}$, a new assignment of Markov blankets to the variables $X_1, \ldots, X_n$ in $\mathcal{H}_q$ is obtained in the following manner. For each X_i and for each q, we compare the value $\sum_{j=1}^{|\mathbf{D}_q|} \log P(x_{i_j}|mb_{q,i_j}(X_i))$ to the value $\sum_{j=1}^{|\mathbf{D}_q|} \log P(x_{i_j}|mb'_{q,i_j}(X_i))$. This values are, respectively, the conditional log-likelihoods of X_i given its Markov blanket, as determined on the one hand by Bayesian network BN_i, and on the other hand by Bayesian network $BN'_{q,i}$. If the latter value is higher than the former, that is if the conditional log-likelihood of X_i given its Markov blanket is increased by replacing $\mathcal{MB}_{q,i}(X_i)$ with $\mathcal{MB}'_{q,i}(X_i)$, then $\mathcal{MB}'_{q,i}(X_i)$ will be the Markov blanket of X_i in $\mathcal{H}_q$ in the new assignment, otherwise X_i will be assigned again $\mathcal{MB}_{q,i}(X_i)$.

2.4   Evaluation Function

An assignment of Markov blankets to the variables is evaluated by measuring the DHRF pseudo-log-likelihood, that is the logarithm of $P^*(O|\mathcal{H})$. Locally (at a specific HRF $\mathcal{H}_q$ level) maximization of the latter is guaranteed by maximization of an evaluation function in the form of a pseudo-log-likelihood of q-th model given the corresponding dataset $\mathbf{D}_q$, namely:

$$\log P^*(\mathbf{D}_q|\mathcal{H}_q) = \sum_{j=1}^{|\mathbf{D}_q|} \sum_{i-1}^{n} \log P(X_i = x_{i_j}|mb_{q,i_j}(X_i)) \qquad (1)$$

Actually, by building an alternative assignment from the current one we implicitly evaluate the new assignment. In fact, the new assignment will differ from the old one only if there is at least one variable X_i such that the new Markov blanket of X_i in $\mathcal{H}_q$ increases the variable conditional log-likelihood given the Markov blanket. Given the definition of the pseudo-log-likelihood function and the way our search operator works, an increase in any one of the n local log-likelihoods of any HRFs (within the DHRF) ensures an increase in the global pseudo-log-likelihood of the DHRF. The reason is clearly that the search operator works in a modular fashion, that is the way each Markov blanket $\mathcal{MB}_{q,i}(X_i)$ is modified by the operator is such that the change does not affect any other Markov blanket in the model. Therefore, after we build a new assignment, it will be sufficient to compare it to the old one in order to know whether it increases the model pseudo-likelihood: if the two assignments are different, then we can endorse the new one as being better, otherwise we keep the old one and stop the search.

2.5   Iteration

Once the structures have been learned, parameters of the DHRF can be estimated as explained in the companion paper [5]. The DHRF trained this way can be used to perform a new, more precise segmentation of the training sequences, building new state-specific training sets $\mathbf{D}_1, \ldots, \mathbf{D}_Q$ from the corresponding subsequences, and the whole structure learning procedure can be carried out all over again in order to obtain improved structures for $\mathcal{H}_1, \ldots, \mathcal{H}_Q$. This iterative scheme completes the dynamic Markov blanket merging algorithm.

3   Empirical Evaluation

In order to evaluate the model empirically, we realized a simulator in JAVA, developed over the *JProGraM* library for HRFs [1]. We compared the DHRF with its natural competitor, the Dynamic Bayesian Network (DBN) (reviewed in short in [5]). The evaluation concerns only synthetic data. In section 3.1 we describe the data generation process, and in section 3.2 we present the results of the experiments.

3.1   Synthetic Data Generation

The data generation process is in two steps: (i) creation of a model which specifies the set of probability laws underlying the data distribution; (ii) generation of the data, drawn from these laws. In practice, we proceed with the generation of a random DHRF, selecting for each random variable in each HRF a random Markov Blanket, and then selecting a random structure for the relative BN. Then, we generate the corresponding, random conditional probability tables (CPTs). Afterwards, we use the resulting DHRF as a probabilistic generative model, drawing the random sequences which constitute our dataset from the joint probability distribution represented by the very DHRF. In order to do that, an inference algorithm is applied, relying on Gibbs sampling as in regular HRFs [2]. Due to the particularities of the present model, inference takes place in two steps. The first step corresponds to the burn-in period. This means that each emission probability model (i.e., each HRF in the DHRF) is initialized at random according to a uniform distribution, and then subjected to an amount of Gibbs iterations that equals the burn-in time. The second step corresponds to the actual sampling phase. The algorithm aims at generating a number of sequences, and each one of them contains observations drawn from the emission probability distributions associated with different latent variables. Hence, for each sequence the initial state is selected and an observation (sampled from the corresponding HRF) is added to the sequence. Then, the next state is determined according to the transition probability distribution, and so on, until a final state is reached. The final state is eventually dismissed, after a certain number of iterations, according to its exit-probability. The core of the process is the generation of individual observations: an observation is sampled from the HRF corresponding to the current latent variable, and it is added to the sequence. Next, the algorithm prescribes to loop over the same latent variable, sampling from the same HRF, for as long as the *intra-sample time*. The latter parameter has to be tuned via model selection techniques, and it defines the expected number of iterations occurring between two samples so that they turn out to be stochastically independent. Then, the next latent variable is generated, and so on. In so doing, sampling from different HRFs is feasible, avoiding problems that may arise from the fact that the number of observations an HRF generates is not known a priori.

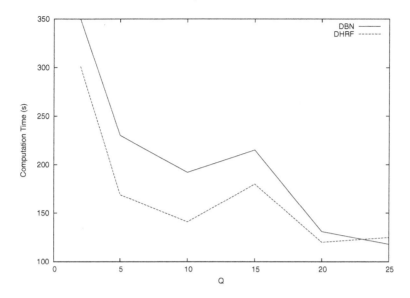

Fig. 1. Computational time of learning for DHRFs and DBNs, respectively, as functions of Q

3.2   Experiments

As we stated in the companion paper [5], the modeling capabilities of DHRFs subsume (by definition, due to the modeling capabilities of HRFs w.r.t. those of BNs) those of DBNs. To all practical ends, it is crucial to compare the performance of DHRFs and DBNs in terms of computational burden on the field. Extension to sequence modeling of the nice computational behavior of static HRFs is one of the main motivations behind the very development of DHRFs in the first place [5]. The focus is thence on the behavior of the learning time with respect to the variations of sensible parameters. These parameters are: (i) the number Q of latent random variables in the model; (ii) the number n of observable random variables in the model. Moreover, statistical (likelihood ratio-like) tests are applied, allowing for a comparison of the fitness of the joint densities estimated via DHRFs and DBNs. The corresponding results are presented hereafter (sections 3.3 and 3.4, respectively). Two different probability distributions, or "classes" ω_1 and ω_2 were generated and used for training/testing the learning machines. In practice, the data drawn from ω_1 and ω_2 were generated by separate generative DHRFs first, and later learned by two class-specific graphical models. In the following, all the results that are not presented in a class-specific fashion are to be intended as overall results, obtained on the whole data sample (inclusive of data drawn from both ω_1 and ω_2). Throughout all the simulations, data generation took place as stated in section 3.1. The data embraced: (i) a training set of S sequences for each class; (ii) a validation set of $S/10$ sequences for each class; (iii) a test set of $S/5$ sequences for each class. In all the simulations we used observable variables whose discrete definition domains had cardinality ranging from 2 to 6 (at random). The experiments were carried out on a computer having PC architecture with CPU Intel Core i5-460M (2.53GHz) and 4GB of RAM.

3.3   Varying the Number of Latent Variables

Here we present a study of the performances of learning algorithms with respect to variations in the number Q of latent variables in the model (i.e. the number of HRFs to be learned). In this scenario we set $n = 20$, $S = 150$, and all sequences have randomly variable length (namely, roughly 100 observations long on average). Remember that in DMBM each HRF H_q is learned on a dataset D_q, obtained via Viterbi segmentation of the sequences in the training set D, as stated in section 2. As a consequence, if Q increases then the expected cardinality of each D_q (for $q = 1, \ldots, Q$) decreases. Hence, our first concern is investigating whether the problem of learning a higher number Q of HRFs leads to an increase in the computation time, or the latter is counterbalanced by the fact that the HRFs have to learn from smaller datasets D_q. This will throw a light on the efficiency of the divide and conquer philosophy underlying DHRFs, a strategy in line with the nature of the bare HRF which turned out to be thriving. Secondly, a direct comparison with DBNs is sought, and accomplished thenceforth.

Table 1.   Pseudo log-likelihood with respect to Q. DHRF $log(P^*)$-t and DHRF $log(P^*)$-D are the pseudo log-likelihoods yielded by the DHRF on the test set and on the training set, respectively. DBN $log(P^*)$-t is the pseudo log-likelihoods yielded by the DBN on the test set. The DHRF ratio is the value $(log(P^*) - t)/(log(P^* - D))$.

Q	DHRF $log(P^*) - t$	DHRF $log(P^*) - D$	DHRF ratio	DBN $log(P^*) - t$
2	-62916	-369528	5,87	-74777
5	-68754	-356400	5,18	-79358
10	-76196	-349941	4,59	-83293
15	-86127	-412663	4,79	-94502
20	-74013	-366310	4,94	-78684
25	-77350	-378012	4,88	-80249

The results of the simulations are shown in figure 1. It is seen that, as Q grows, the computational burden of DHRFs is relatively stable. Yet, its predominant trend is even slowly decreasing. This suggests the preliminary conclusion that the computational complexity caused by the introduction of additional latent variables is indeed counterbalanced by the proportional decrease in the cardinality of the segmented datasets D_q. This consideration may encourage the practitioner to apply DHRFs including a large number of latent variables, since an increase in expressiveness of the model may be expected without any loss in terms of complexity. Unfortunately, this may not be taken as a general rule of the thumb. In fact, as shown by the values of the *DHRF ratio* in table 1, an increase of Q over a certain threshold leads to a poor generalization capability. The DHRF ratio is defined as $(log(P^*) - t)/(log(P^*) - D)$, that is clearly an expression of the generalization capability of the learning machine. Of course, this phenomenon is just an instance, in the present scenario, of the traditional bias-variance dilemma (or, of the over-learning occurring as a consequence of the implicit increase in the Vapnik-Chervonenkis dimension of the resulting DHRF). Furthermore, in case the forward-backward algorithm for parameter learning [5] is applied afterwards, the overall time complexity would be affected by Q eventually (since in that variant of the algorithm the dataset D does not undergo initial segmentation, so that each one of the Q HRFs in the DHRF would learn from the whole D). Established model selection techniques are thus required from time to time in order to come up with a value for Q which fits the data at hand.

Finally, it is noteworthy that the results shown in table 1 highlight the significant difference between the $log(P^*)$ yielded by the DHRF and by the DBN, respectively. The former model is able to capture more statistical information (i.e., more conditional (in)dependences) from the training data than the latter one in a (at least) tantamount time, as expected.

3.4   Varying the Number of Observable Variables

Next, we analyze the performance of the learning algorithms with respect to variations in the number n of observable variables. Here we set $Q = 5$, $S = 200$, and random, variable sequence length (roughly 50 observations per sequence, on average). The present experiment is of the utmost relevance to our goals, since n is the most sensible parameter in relation to both the computational efficiency of learning in graphical models, and the capability to fit real-world scenarios that involve a number of random quantities (or, features). Studies presented in [3,4] show that the HRF scales very well with respect to n, while generally BNs and Markov random fields do not. As for DHRFs for sequences, the results of the simulations are reported in tables 2 and 3. The computational efficiency of DHRFs with DMBM in modeling large sets of random variables is immediately seen from figure 2. Table 2 reports also on the experiments we carried out with $n = 80$ and $n = 100$, where DBNs could not complete their learning process within an acceptable deal of time (i.e., the corresponding experiments had to be terminated). Basically, the computation time appears to increase (with n) in an approximately linear way, while the complexity of the learning algorithms for DBNs increase roughly exponentially.

Next, we investigate the variations of the pseudo log-likelihood $log(P^*)$ yielded by the two models on the test set. As shown in figure 3, the DHRF scores significantly higher than the DBN. Table 3 reports on the corresponding values

Table 2. Learning time in minutes and seconds (min:sec) as a function of n. The values for $n = 80$ and $n = 100$ could not be computed for the DBN.

n	DHRF time	DBN time
10	2:15	0:39
20	3:26	5:07
30	6:59	31:28
40	5:46	83:34
50	9:09	192:57
60	15:12	712:10
80	21:45	-
100	27:57	-

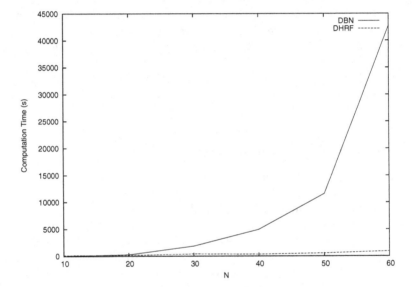

Fig. 2. Graphical comparison of the learning time for DHRF and DBN, respectively, as functions of n

Table 3. Pseudo likelihood ratio (PL-ratio) obtained over the data distributions ω_1 and ω_2 as a function of n

n	PL-ratio for ω_1	PL-ratio for ω_2
10	4834	13042
20	10472	16118
30	21394	29286
40	22708	17896
50	33730	29080
60	50916	36454

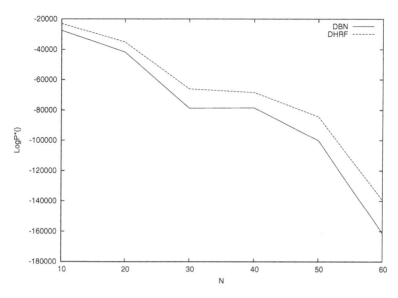

Fig. 3. Graphical comparison of the pseudo log-likelihoods yielded by the two models as functions of n

of pseudo likelihood ratio (PL-ratio) between the two models for the separate populations ω_1 and ω_2, respectively. The PL-ratio is defined as:

$$
\begin{aligned}
PL - ratio &= -2ln\left\{\frac{P^*_{DBN}(Testset)}{P^*_{DHRF}(Testset)}\right\} \\
&= -2ln(P^*_{DBN}(Testset)) + 2ln(P^*_{DHRF}(Testset))
\end{aligned}
\tag{2}
$$

This is a proper instance of the classic likelihood-ratio statistical test. In practice, positive values of the pseudo likelihood ratio entail the statistically-grounded fact that the DHRF fits the data better than the DBN does. As seen from table 3, we observed $PL - ratio \gg 0$ in a systematic manner throughout the simulations. Put in words, the experimental evidence shows that (in the present scenario) the DHRF exhibits an improved behavior over the DBN both in terms of the capability to fit the data, and of the corresponding computational burden.

4   Conclusion

The paper introduced a technique for structure learning in DHRFs. The technique turns out to be a viable method for capturing the statistical (in)dependencies among the random variables within a sequence. Complexity issues were tackled by means of adequate strategies from classic literature on probabilistic graphical models. Performance of the resulting learning machine, suitable for the classification of sequential patterns, was empirically evaluated in a synthetic task. Preliminary results show that DHRFs compare favorably with DBNs in terms of computational burden, with no loss in terms of modeling capability.

Acknowledgments. The invaluable support by Antonino Freno is gratefully acknowledged.

References

1. Freno, A.: JProGraM - PRObabilistic GRAphical Models in Java (2009), http://www.dii.unisi.it/~freno/JProGraM.html
2. Freno, A., Trentin, E.: Hybrid Random Fields: A Scalable Approach to Structure and Parameter Learning in Probabilistic Graphical Models. Springer (2011)
3. Freno, A., Trentin, E., Gori, M.: Scalable Pseudo-Likelihood Estimation in Hybrid Random Fields. In: Elder, J.F., Fogelman-Souli, F., Flach, P., Zaki, M. (eds.) Proceedings of the 15th ACM SIGKDD Conference on Knowledge Discovery and Data Mining (KDD 2009), pp. 319–327. ACM (2009)
4. Freno, A., Trentin, E., Gori, M.: Scalable Statistical Learning: A Modular Bayesian/Markov Network Approach. In: Proceedings of the International Joint Conference on Neural Networks (IJCNN 2009), pp. 890–897. IEEE (2009)
5. Trentin, E., Bongini, M.: Towards a Novel Probabilistic Graphical Model of Sequential Data: Fundamental Notions and a Solution to the Problem of Parameter Learning. In: Mana, N., Schwenker, F., Trentin, E. (eds.) ANNPR 2012. LNCS (LNAI), vol. 7477, pp. 72–81. Springer, Heidelberg (2012)

Statistical Recognition of a Set of Patterns Using Novel Probability Neural Network

Andrey V. Savchenko

National Research University Higher School of Economics, Nizhniy Novgorod,
Russian Federation
avsavchenko@hse.ru

Abstract. Since the works by Specht, the probabilistic neural networks (PNNs) have attracted researchers due to their ability to increase training speed and their equivalence to the optimal Bayesian decision of classification task. However, it is known that the PNN's conventional implementation is not optimal in statistical recognition of a set of patterns. In this article we present the novel modification of the PNN and prove that it is optimal in this task with general assumptions of the Bayes classifier. The modification is based on a reduction of recognition task to homogeneity testing problem. In the experiment we examine a problem of authorship attribution of Russian texts. Our results support the statement that the proposed network provides better accuracy and is much more resistant to change the smoothing parameter of Gaussian kernel function in comparison with the original PNN.

Keywords: Statistical pattern recognition, sets of patterns, probabilistic neural network, hypothesis test for samples homogeneity.

1   Introduction

Pattern recognition [1] is a fundamental aspect of many tasks in artificial intelligence, data mining, computer vision, medical diagnostics, decision-support systems. These tasks may be formulated [2] in terms of statistical recognition [3], [4] of a set of patterns: it is required to estimate the class of an input sample of random variables, with an assumption that all available information about each class is concluded in certain samples of observations [2]. This general formulation could be applied to such acute tasks as image recognition, voice phonemes recognition, authorship attribution, etc.

This problem is usually reduced [5] to a statistical classification of the query sample. The optimal decision is taken with a minimum Bayes risk principle [3]. The unknown probability density, required in this approach, is usually estimated by means of nonparametric techniques [6], [7] like kernel discriminant analysis, e.g. Parzen approach [8]. Such estimations were proved to converge to the real probability density if the training sample size is large [9], [10]. The widely-used parallel implementation of nonparametric approach is a probabilistic neural network (PNN) [11]. This

N. Mana, F. Schwenker, and E. Trentin (Eds.): ANNPR 2012, LNAI 7477, pp. 93–103, 2012.

multilayered feedforward network introduced by Specht [12], [13] is characterized by extremely fast training procedure.

The PNN was proved to be an asymptotically-optimal rule in the classification task [11], [14] if a query object is a single feature vector. Unfortunately, conventional PNN does not provide an optimal solution [2] if the query object is represented by a set of features with the size approximately equal to the training set size. Really, in this case the task should be reduced to a homogeneity testing of query and training samples [15]. In this paper we introduce the modification of the PNN, which saves all advantages of the conventional PNN but yields an optimal decision boundary in statistical recognition of a set of patterns. We experimentally show that the proposed PNN achieves better accuracy and is much more resistant to change the smoothing parameter of Gaussian kernel function [16].

The rest of the paper is organized as follows: Section 2 presents statistical recognition of a set of patterns using the PNN [11]. In Section 3, we introduce our PNN modification. In Section 4, we present the experimental results in the author identification task [17] with well-known texts from Russian literature [18]. Finally, concluding comments are given in Section 5.

2   Statistical Recognition of a Set of Patterns

Let a set $\mathbf{X} = \{\mathbf{x}_j\}, j = \overline{1,n}$ of independent identically distributed (i.i.d.) random variables with unknown $\mathbf{P}$ probability distribution be specified. Here n is a sample size, $\mathbf{x}_j = \{x_{j;1}, ..., x_{j;M}\}$ - is a vector of features with a fixed dimension $M = const$. The pattern recognition problem is to estimate the class of $\mathbf{X}$. It is assumed that each class $r \in \{1.,,.R\}$ is defined by a training set of i.i.d. random variables with unknown $\mathbf{P}_r$ probability distribution $\mathbf{X}_r = \left\{\mathbf{x}_j^{(r)}\right\}, j = \overline{1,n_r}$. Here n_r is a training sample size, $\mathbf{x}_j^{(r)}$ is a feature vector with dimension M.

Following the statistical approach [2], [3], we assume that each class is fully determined by the distribution $\mathbf{P}_r, r = \overline{1,R}$ of its feature vector. Thus, the problem is referred to a hypothesis testing for distribution of $\mathbf{X}$

$$W_r: \qquad \mathbf{P} = \mathbf{P}_r \quad r = \overline{1,R} \tag{1}$$

To solve the problem (1), the principle of minimum Bayes risk [4] is applied. The query sample $\mathbf{X}$ is assigned to the class v with maximum a-posterior probability

$$v = \underset{r \in \{1,...,R\}}{\arg\max} \; P(\mathbf{X}|W_r) \cdot P(W_r) \tag{2}$$

Here $P(W_r)$ is the prior class probability, $P(\mathbf{X}|W_r)$ is a conditional class density (likelihood). In the most practically important pattern recognition tasks [19], [20] it is assumed that each class is equiprobable (prior uncertainty): $P(W_r) = \dfrac{1}{R}$. Following the nonparametric approach, the likelihood is estimated by the given training set with a kernel trick [21]

$$\hat{P}(\mathbf{X}|W_r) = \frac{1}{(n_r)^n} \prod_{j=1}^{n} \sum_{j_r=1}^{n_r} K_{n_r}\left(\mathbf{x}_j, \mathbf{x}_{j_r}^{(r)}\right) \tag{3}$$

Here $K_{n_r}\left(\mathbf{x}_j, \mathbf{x}_{j_r}^{(r)}\right)$ is a kernel function [14]. For example, the Gaussian Parzen kernel [8] is widely used [11]

$$K_{n_r}\left(\mathbf{x}_j, \mathbf{x}_{j_r}^{(r)}\right) = \frac{1}{\left(2\pi\sigma^2\right)^{M/2}} \exp\left(-\frac{1}{2\sigma^2} \sum_{i=1}^{M}\left(x_{j;i} - x_{j_r;i}\right)^2\right) \tag{4}$$

Here $\sigma = const > 0$ is a fixed smoothing parameter (standard deviation of the Gaussian kernel). Based on the estimate (3), the final decision (2) could be written as

$$v = \arg\max_{r \in \{1,...,R\}} \frac{1}{(n_r)^n} \prod_{j=1}^{n} \sum_{j_r=1}^{n_r} K_{n_r}\left(\mathbf{x}_j, \mathbf{x}_{j_r}^{(r)}\right) \tag{5}$$

The criterion (5) corresponds the PNN for statistical recognition of a set of patterns problem (1). In contrast with the conventional four-layered PNN [11], [12], which is used to classify one object $\mathbf{x}_j$, the network (5) contains additional, production, layer to classify the sample of objects $\mathbf{X}$.

3   An Optimal Algorithm

Pattern recognition problem is characterized by the unknown probability distributions $\mathbf{P}_r$ of each class r. Thus, it is required to estimate the conditional density (3). It is the key difference [2] from the conventional classification task, in which distributions $\mathbf{P}_r$ are given. Hence, we believe it is better to follow the Borovkov's approach [2] rather than refer pattern recognition task to a statistical testing for distribution (1). According to this approach [2], the problem is reduced to a testing for homogeneity of input sample $\mathbf{X}$ and training set $\mathbf{X}_r$:

$W_r : \left\{ \mathbf{x}_j \right\}, j = \overline{1,n}$ and $\left\{ \mathbf{x}_j^{(r)} \right\}, j = \overline{1,n_r}$ have the same probability density

The decision is made with a minimum Bayes risk principle by using the set from the united sample space $\{\mathbf{X}, \mathbf{X}_1, ..., \mathbf{X}_R\}$ to a class v

$$v = \underset{r \in \{1,...,R\}}{\arg\max} \; \underset{\mathbf{P}^* \; \mathbf{P}_j^*, \, j = \overline{1,R}}{\sup} \; P\left(\{\mathbf{X}, \mathbf{X}_1, ..., \mathbf{X}_R\} | W_r\right) \cdot P(W_r) \tag{6}$$

Here $\mathbf{P}^*$ is a possible probability distribution of a sample $\mathbf{X}$, $\mathbf{P}_j^*$ is a possible distribution of jth training sample. Assuming the independence of random variables in a united sample $\{\mathbf{X}, \mathbf{X}_1, ..., \mathbf{X}_R\}$, the conditional density in (6) could be written as

$$\underset{\mathbf{P}^* \; \mathbf{P}_j^*, \, j = \overline{1,R}}{\sup} \; P\left(\{\mathbf{X}, \mathbf{X}_1, ..., \mathbf{X}_R\} | W_r\right) =$$

$$= \frac{\underset{\mathbf{P}^*}{\sup} P(\mathbf{X}|W_r) \underset{\mathbf{P}_r^*}{\sup} P(\mathbf{X}_r|W_r)}{\underset{\mathbf{P}_r^*}{\sup} P(\mathbf{X}_r)} \cdot \prod_{j=1}^{R} \underset{\mathbf{P}_j^*}{\sup} P(\mathbf{X}_j)$$

As $\prod_{j=1}^{R} \underset{\mathbf{P}_j^*}{\sup} P(\mathbf{X}_j) = const(\mathbf{X})$ does not depend on a query sample $\mathbf{X}$, and $\underset{\mathbf{P}^*}{\sup} P(\mathbf{X})$ is independent on $\mathbf{X}_r$, we convert (6) to the following expression

$$v = \underset{r \in \{1,...,R\}}{\arg\max} \; \frac{\underset{\mathbf{P}^*}{\sup} P(\mathbf{X}|W_r) \underset{\mathbf{P}_r^*}{\sup} P(\mathbf{X}_r|W_r)}{\underset{\mathbf{P}^*}{\sup} P(\mathbf{X}) \cdot \underset{\mathbf{P}_r^*}{\sup} P(\mathbf{X}_r)} \cdot P(W_r) \tag{7}$$

It is known [2], that the supremum of the likelihood is achieved when the valid probability distributions $\mathbf{P}^*$, $\mathbf{P}_r^*$ are equal to their optimal unbiased estimates. Herewith, to evaluate this optimal estimate the combined sample $\{\mathbf{X}, \mathbf{X}_r\}$ is used if the condition W_r is true, i.e. $\mathbf{X}$ and $\mathbf{X}_r$ are the samples of the same random variable. Hence, we may use nonparametric kernel estimation (3), e.g.

$$\sup_{\mathbf{P}^*} P(\mathbf{X}|W_r) =$$

$$= \frac{1}{(n+n_r)^n} \prod_{j=1}^{n} \left(\sum_{j_1=1}^{n} K_n\left(\mathbf{x}_j, \mathbf{x}_{j_1}\right) + \sum_{j_r=1}^{n_r} K_{n_r}\left(\mathbf{x}_j, \mathbf{x}_{j_r}^{(r)}\right) \right),$$

$$\sup_{\mathbf{P}^*} P(\mathbf{X}) = \frac{1}{n^n} \prod_{j=1}^{n} \sum_{j_1=1}^{n} K_n\left(\mathbf{x}_j, \mathbf{x}_{j_1}\right).$$

Thus, (7) is equivalent to the following criterion

$$v = \arg\max_{r \in \{1,...,R\}} \frac{n^n \cdot (n_r)^{n_r}}{(n+n_r)^{n+n_r}} \prod_{j=1}^{n} \left(1 + \frac{\sum_{j_r=1}^{n_r} K_{n_r}\left(\mathbf{x}_j, \mathbf{x}_{j_r}^{(r)}\right)}{\sum_{j_1=1}^{n} K_n\left(\mathbf{x}_j, \mathbf{x}_{j_1}\right)} \right) \times$$

$$\times \prod_{j_r=1}^{n_r} \left(1 + \frac{\sum_{j_1=1}^{n} K_n\left(\mathbf{x}_{j_r}^{(r)}, \mathbf{x}_{j_1}\right)}{\sum_{j_{r;1}=1}^{n_r} K_{n_r}\left(\mathbf{x}_{j_r}^{(r)}, \mathbf{x}_{j_{r;1}}^{(r)}\right)} \right)$$

$$(8)$$

Expression (8) corresponds to a proposed PNN for recognition of a set of patterns. Its implementation is shown in Fig. 1. Here the input layer contains not only the query sample $\mathbf{X}$, but the united sample $\{\mathbf{X}, \mathbf{X}_1,..., \mathbf{X}_R\}$. The kernel function for a query sample is added to a training set in the second, pattern, layer. The new division layer is added according to (8). In the production layer we multiply not only the features of the query object $\mathbf{X}$, but also features of rth sample $\mathbf{X}_r$.

It could be noticed that if $n_r \to \infty$, expression (8) is equivalent to (5). Really, in asymptotics, the training set $\mathbf{X}_r$ fully determines the probability distribution $\mathbf{P}_r$. Hence, the united sample $\{\mathbf{X}, \mathbf{X}_r\}$ does not provide any additional information.

The proposed PNN saves all advantages of the classical PNN [11], but the rate of convergence to the optimal decision should be higher for (8) than for a classical implementation (5). The next section provides an experimental evidence to support this claim.

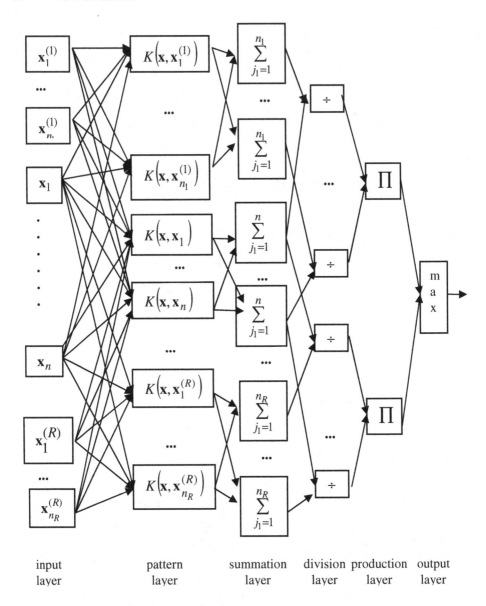

Fig. 1. Proposed modification of the PNN in statistical recognition of a set of patterns

4 Experimental Results

In this section we demonstrate the proposed modification of PNN (8) in the pattern recognition problem from the linguistic analysis. It is required to identify the author of a Russian text fragment [17], [22]. The training sets contain other text extracts. We use the following eight well-known large Russian texts: "Anna Karenina" and "Resurrection" by L. Tolstoy, "Idiot" and "Crime and punishment" by F. Dostoevsky, "Dead souls" and "Taras Bulba" by N. Gogol, "And Quiet Flows the Don" and "Virgin Soil Upturned" by M. Sholokhov. All texts (in original) were taken from the corpus [18].

We compare the accuracy of the proposed PNN (8) with its conventional implementation (5). Additionally, we provide error rate for conventional multilayer perceptron (MLP) with one hidden layer trained by using the backpropagation. To classify the sample of objects $\mathbf{X}$, we use production (5) of MLP outputs for a single pattern $\mathbf{x}_j$. All neural networks were implemented as a parallel application with Java Runtime Environment 1.7 on a modern laptop (Intel Core i7 CPU 2.0 GHz, 6 Gb RAM).

The accuracy was estimated by the following procedure. One randomly chosen fragment of each text with fix size (in characters) was added to the training set. The test set contain 10 randomly chosen fragments of each text (i.e., 8*10=80 samples). The error rate was estimated by these 80 tests. The total recognition quality was estimated as an arithmetical mean of the error rate by 100 such experiments of training and test set selection (i.e., 80*100=8000 classifications).

In the first experiment we used the frequency of punctuation marks in each sentence as a simple but informative feature of Russian language. This feature set is known to show a good quality in Russian authorship attribution [17]. The following punctuation marks were chosen: ".", "?", "!", ",", ".", "-", ":", ";", "("; i.e. the feature vector $\mathbf{x}$ contains $M=9$ features. In the first case both training and test sets were generated by fragments of 25000 characters (the sample size $n=900...1150$).

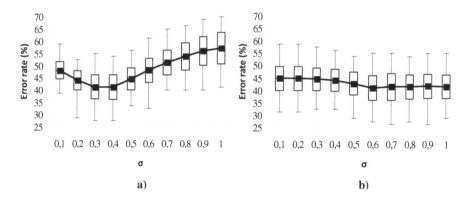

Fig. 2. Dependence of average error rate (in %) on σ for 25000 characters of each fragment both in training and test sets. a. criterion (5); b. criterion (8).

The box-plot diagrams of dependence of classification error rate on the smoothing parameter σ of Gaussian kernel function (4) are shown in Fig. 2. For comparison, the best error rate for MLP was achieved with 100 neurons in a hidden layer and is equal to 58,3%±20%.

The average recognition time per one set t_{rec} here is equal to 0.1s., 0.25s. and 0.4s. for the original PNN (5), proposed criterion (8) and MLP, respectively. The average training time t_{tr} is equal to 0.01s. for both (5), (8) and 34s for MLP.

As we could see in Fig.2, the minimal error rate of criterion (8) is equal to 40.8% , which is a bit less then the minimal error rate 41.3% of the conventional PNN (5). The most significant quality indicator of proposed network (Fig. 1) is the robustness of error rate dependence on the smoothing parameter. Really, the error rate for the proposed criterion (8) is always less than 45%. At the same time, accuracy of the traditional PNN varies enormously. At worst, the average error rate is equal to 59%.

In the second case the experiment was repeated, but the training sets were generated from 100000 characters in a fragment (i.e. the sample size $n=3400...4200$). The test set was still generated by 25000 character-fragments. The results are illustrated with a box-plot diagram in Fig. 3. The best error rate for MLP was achieved here with 350 neurons in a hidden layer and is equal to 52,2%±23,4%.

The average recognition time per one set t_{rec} is equal to 0.5s., 1.1s. and 0.9 s. for the original PNN (5), proposed criterion (8) and MLP respectively. The average training time t_{tr} is equal to 0.05s. for both (5), (8) and 920s. for MLP.

In Fig. 3 one could see that the recognition accuracy is extremely better in comparison with the previous experiment. The minimal average error rate (23%) of (8) is a bit less than the minimal error rate 23.6% of the PNN (5). Again, the proposed network is more robust to change the smoothing parameter than the traditional PNN.

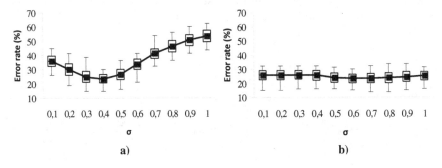

Fig. 3. Dependence of average error rate (in %) on σ for 100000 characters of fragment in training set and 25000 characters of fragment in test sets. a. criterion (5); b. criterion (8).

In the last experiment we chose the frequency of the bigram of words [17]. This feature is widely used in natural language processing and showed good quality of author attribution [22]. We extract 500 most frequent word from the union of all training fragments and calculate the frequency of each word from each fragment. As a preliminary processing we perform stemming, i.e. removing the known Russian endings. Words beginning with capitals were omitted (including the first words of sentences). Finally, we retain only words with length greater than 3 characters. The frequency of the same 500 words were calculated for each test fragment. The kernel function in this experiment is a limit of Gaussian kernel (4) when $\sigma \to 0$, i.e. the discrete delta function. Here the author identification error rate of the proposed PNN (8) is 13.3% if the training set was generated from 25000 character-fragments. It is much less than the error rate of the conventional PNN (27.8%). The best recognition results for different number of characters in a training text fragments is summarized in Table 1.

Table 1. The average error rate of the author identification by text fragment

Features	Frequency of punctuation in a sentence		Frequency of word bigram	
Number of characters	Criterion (5)	Criterion (8)	Criterion (5)	Criterion (8)
25000 - training set, 25000 - test set	41,3%±7,5%	40,8%±7,4%	26,4%±8,3%	13,1%±6,9%
100000 - training set, 25000 - test set	23,6%±4,5%	23,0%±4,4%	19,8%±5,7%	3,6%±3,0%

Based on this results we could draw the following conclusion. First, the classification accuracy of the conventional PNN is not greater than the accuracy of the proposed modification (8). It is particularly noticeable for recognition with discrete features and discrete delta function as a kernel. Second, the task of proper choice of the best value of the smoothing parameter σ of Gaussian kernel (4) for proposed criterion (8) is not as acute as for traditional criterion (5). Third, the author identification quality of synthesized criterion (8) with frequency of words as a feature (the last column of Table 1) is rather good even in comparison with known best results for Russian texts [17]. And, fourth, conventional MLP is not the best choice in this task because model sets of objects may contain equal patterns. As a matter of fact, MLP should be used to compare patterns extracted from the model sets. such as an estimation of distribution (3). Unfortunately, this procedure shows good recognition quality only if the database contains several samples per class. In our experiment we have one class per sample, hence this approach has not been applied.

5   Conclusion

The proposed network (Fig. 1) is a generalization of the conventional PNN in statistical recognition problem of a set of patterns. Our modification has all known advantages of the PNN [11] over classifiers based on other neural networks. First of all, it is an excellent training speed in comparison with back propagation. The new sample can be added even in real time applications. The network begins to generalize each new observed set of patterns causing the decision boundary to become closer to the optimal one. Unlike many networks, the PNN (8) does not contain recursive connections from the neurons back to the inputs. Thus, it could be implemented completely in parallel [12].

The key difference of (8) from other approaches is the usage of the query sample $\mathbf{X}$ to estimate the joint probabilistic quantity of the united sample $\{\mathbf{X}, \mathbf{X}_1, ..., \mathbf{X}_R\}$. Hence, the distribution of each model class r is estimated by the united sample $\{\mathbf{X}, \mathbf{X}_r\}$. The query sample is the part of the second, pattern, layer of the PNN, and each model sample became a member of the first, input, layer.

The most significant advantage of the proposed classifier (8) is that its decision boundary converges to the Bayes optimal solution [2]. The rate of the convergence is essentially higher than the rate for the classical PNN (5) in the most acute case of a pattern recognition when the training sample size is approximately equal to the size of an input sample $n_r \approx n$. At last, the proper choice of the smoothing parameter σ of the Gaussian kernel (4) is not as complex as for the conventional PNN [16], [23]. Our experimental study showed that the criterion (8) is much more resistant to change σ than the PNN (5), though the maximal accuracy of (8) is practically equal to the maximal accuracy of (5). Thus, our network (Fig. 1) could achieve better quality in time-varying environment. However, the accuracy of the proposed network is 2-5 times better than the accuracy of the PNN (5) with a discrete delta function kernel (8), which is known to be a limit ($\sigma \to 0$) of Gaussian kernel (4).

One other advantage of the proposed PNN is that the measure of similarity in (8) is symmetric as the importance of training and input sets is equivalent in the homogeneity testing. On the other hand, the similarity measure in (5) is asymmetric. Really, in statistical classification the model probability distribution (evaluated by the training sample) is much more important than the input sample distribution. In this task it is supposed that $n_r \gg n$, so the quality of the model probability distribution estimation is much higher than the quality of the query sample. This fact is an additional argument in behalf of (8) as symmetry is a desired property in many pattern recognition algorithms [1] (e.g., clustering).

Unfortunately, our network (Fig. 1) possesses the same shortcoming as the PNN. First of all, our network requires large memory to store all training samples. Second, the classification speed is low as the network is based on an exhaustive search through all training samples [20]. Moreover, the proposed network classifies the input sample twice slower than the original PNN. However, this fact is not a real obstacle in practical pattern recognition tasks (author identification, image recognition), as the training sample size is not usually large. Third, our network is not as general as the traditional PNN because we require the network input to be a sample with the size which is the same order of magnitude as the training sample size [1], [20].

Thus, in this study we proposed the novel modification of the PNN (8) in recognition of a set of patterns. We experimentally proved that this network is in some terms better than the conventional PNN (5). The PNN modification proposed here can be used in various pattern recognition tasks [1], such as image and speech recognition.

Acknowledgements. This research is supported by Federal Grant-in-Aid Program "Research and development on priority directions of scientific-technological complex of Russia for 2007-2013" (Governmental Contract No. 07.514.11.4137).

References

1. Theodoridis, S., Koutroumbas, C.: Pattern Recognition, 4th edn. Elsevier Inc. (2009)
2. Borovkov, A.A.: Mathematical Statistics. Gordon and Breach Science Publishers (1998)
3. Vapnik, V.N.: Statistical Learning Theory. Wiley, New York (1998)
4. Webb, A.R.: Statistical Pattern Recognition. Wiley, New York (2002)
5. Duda, R.O., Hart, P.E., Stork, D.G.: Pattern Classification. Wiley, New York (2001)
6. Efromovich, S.: Nonparametric Curve Estimation. Methods, Theory and Applications. Springer, New York (1999)
7. Murthy, V.K.: Estimation of probability density. Annals of Mathematical Statistics 36, 1027–1031 (1965)
8. Parzen, E.: On estimation of a probability density function and mode. Annals of Mathematical Statistics 33, 1065–1076 (1962)
9. Greblicki, W.: Asymptotically optimal pattern recognition procedures with density estimates. IEEE Transactions on Information Theory IT-24, 250–251 (1978)
10. Wolverton, C.T., Wagner, T.J.: Asymptotically optimal discriminant functions for pattern classification. IEEE Transactions on Information Theory 15, 258–265 (1969)
11. Specht, D.F.: Probabilistic neural networks. Neural Networks 3, 109–118 (1990)
12. Specht, D.F.: Probabilistic Neural Networks for Classification, Mapping, or Associative Memory. In: IEEE International Conference on Neural Networks, vol. I, pp. 525–532 (1988)
13. Specht, D.F.: A general regression neural network. IEEE Transactions on Neural Networks 2(6), 568–576 (1991)
14. Rutkowski, L.: Adaptive Probabilistic Neural Networks for Pattern Classification in Time-Varying Environment. IEEE Transactions on Neural Networks 15(4), 811–827 (2004)
15. Kullback, S.: Information Theory and Statistics. Dover Pub. (1997)
16. Jones, M.C., Marron, J.S., Sheather, S.J.: A brief survey of bandwidh selection for density estimation. Journal of the American Statistical Association 91, 401–407 (1996)
17. Kukushkina, O.V., Polikarpov, A.A., Khmelev, D.V.: Using Literal and Grammatical Statistics for Authorship Attribution. Problems of Information Transmission 37(2), 172–184 (2001)
18. The e-library of Maxim Moshkov, http://www.lib.ru
19. Savchenko, A.V.: Image Recognition with a Large Database Using Method of Directed Enumeration Alternatives Modification. In: Kuznetsov, S.O., Ślęzak, D., Hepting, D.H., Mirkin, B.G. (eds.) RSFDGrC 2011. LNCS (LNAI), vol. 6743, pp. 338–341. Springer, Heidelberg (2011)
20. Savchenko, A.V.: Directed enumeration method in image recognition. Pattern Recognition 45(8), 2952–2961 (2012)
21. Aizerman, M.A., Braverman, E.M., Rozonoer, L.I.: Theoretical foundations of the potential function method in pattern recognition learning. Automation and Remote Control 25, 821–837 (1964)
22. Stamatatos, E.: A survey of modern authorship attribution methods. Journal of the American Society for Information Science and Technology 60(3), 538–556 (2009)
23. Mao, K.Z., Tan, K.-C., Ser, W.: Probabilistic neural-network structure determination for pattern classification. IEEE Transactions on Neural Networks 11, 1009–1016 (2000)

On Graph-Associated Matrices and Their Eigenvalues for Optical Character Recognition

Miriam Schmidt, Günther Palm, and Friedhelm Schwenker

Institute of Neural Information Processing,
University of Ulm, 89069 Ulm, Germany
{miriam.k.schmidt,guenther.palm,friedhelm.schwenker}@uni-ulm.de
http://www.uni-ulm.de/in/neuroinformatik.html

Abstract. In this paper, the classification power of the eigenvalues of six graph-associated matrices is investigated and evaluated on a benchmark dataset for optical character recognition. The extracted eigenvalues were utilized as feature vectors for multi-class classification using support vector machines. Each graph-associated matrix contains a certain type of geometric/spacial information, which may be important for the classification process. Classification results are presented for all six feature types, as well as for classifier combinations at decision level. For the decision level combination probabilistic output support vector machines have been applied. The eigenvalues of the weighted adjacency matrix provided the best classification rate of 89.9 %. Here, almost half of the misclassified letters are confusion pairs, such as *I-L* and *N-Z*. This classification performance can be increased by decision fusion, using the sum rule, to 92.4 %.

Keywords: graph classification, weighted adjacency matrix, spectrum, support vector machine.

1 Introduction

Spectral graph theory is an important branch in the area of graph classification. Matrices associated with graphs, e.g. the adjacency matrices, contain essential information about the graph's connectivity [1]. This information is also included in the eigenvalues of the adjacency matrix which build the so-called spectrum. The spectrum of a graph exhibits some important properties, which make them ideal candidates for classification tasks [2],[3].

First, the spectrum is invariant with respect to the labeling of the nodes. Two graphs, which only differ in the labeling, are called isomorph to each other and the graph isomorphism problem (the computation, if two graphs are isomorph) belongs to the NP-complete problems. These problems build a subset of NP problems, which are defined as decision problems whose solutions can be verified in polynomial time, but the time required to solve the problems increases quickly [4]. The graph isomorphism problem occurs, if one has to match the nodes of two graphs to calculate, for example, the graph edit distance [5],[6].

N. Mana, F. Schwenker, and E. Trentin (Eds.): ANNPR 2012, LNAI 7477, pp. 104–114, 2012.
© Springer-Verlag Berlin Heidelberg 2012

Hereby, a distance between two graphs is calculated, taking the number and corresponding costs of operations (deleting or inserting nodes or edges, relabeling) into account which are needed to transform one graph into the other. By using the spectrum as feature of the graph, it is not necessary to deal with this matching problem. However, if two graphs are just flipped or rotated, the spectrum does not provide the ability to distinguish these isomorph graphs.

Furthermore, if the underlying matrix is real and symmetric, the eigenvalues are also real. Hence, the eigenvalues can be used to map the graphs in a coordinate system and use well known clustering or classification algorithms. This method is called spectral embedding [7],[8].

In some applications, graphical representations of different orders have to be considered. In the underlying data set (see Sect. 4) are graphs with one node up to graphs with eight nodes. This can be easily accomplished by using only a distinct subset of the eigenvalues, e.g. the first three eigenvalues [9].

But one big restriction to the adjacency matrix is that it includes no information about the length of the edges in the graph. It contains just the binary information, if there is an edge or not. If two graphs have the same edges, but differ in the stretching of the graph, it is impossible to distinguish them by using the adjacency matrices, because they are exactly the same. Thus, to include the information about the stretching, the weighted adjacency matrices (with the lengths of the edges as labels) can be used [10]. In this case, the binary values in the matrices are replaced by the labels of the corresponding edges.

In this paper, the power of the principal eigenvalues of six different graph-associated matrices for the spectral classification is investigated. Each matrix ((weighted) adjacency matrix, (weighted) Laplacian matrix and (weighted) adjacency matrix of the complement graph) includes different information, which can be crucial for the classification or needless. Classification results are presented for all six feature types and by investigating the miss-classified samples, we were able to differentiate the qualities of the different matrices. To improve the classification performance, we also utilized classifier combinations at decision level. For the decision level combination probabilistic output support vector machines habe been applied.

As application data set, we used the capital letter data set of the *IAM Graph Database Repository*[1] because of its publicity, its complexity and its accessibility. Our classification results are below the best performance on this dataset [11], but the aim was not to outperform the existing classification approaches but to investigate different spectra for classification problems.

The rest of the paper is organized as follows: First, in Sect. 2 the essential notations in (spectral) graph theory are provided. The illustration of the support vector machine classifier in Sect. 3 is followed by the data description including the explanation of the feature extraction in Sect. 4. In Sect. 5 the experiments and the results are presented before the paper will be completed by a summary and conclusion.

---

[1] Databases of the Institute of Computer Science and Applied Mathematics in Bern, Switzerland. http://www.iam.unibe.ch/fki/databases/iam-graph-database

2   Graph Associated Matrices

This section provides a brief introduction in graph theory. The textbooks [12] and [13] are recommended for extensive information about these topics.

An *undirected graph* is defined as a pair $G = (V, E)$, consisting of a finite set of nodes $V(G) = \{v_1, ..., v_n\}$ and a set of edges $E(G) = \{e_1, ..., e_m\}$, which are unordered two-element subsets of $V(G)$ with $e = \{v_k, v_l\}$, v_k, v_l in $V(G)$. The order of a graph is described by $|V(G)| = n$, the size by $|E(G)| = m$. The degree of v_k is $d(v_k) = |\{v_l \in V(G)|\{v_k, v_l\} \in E(G)\}|$, $k, l = 1, ..., n$ (the number of edges connected with v_k).

A graph $G = (V, E, w)$ is called *weighted graph*, with a labeling function $w(e_i)$, which assigns a label to each edge e_i. Usually real numbers are used as labels (e.g. length, cost).

The most known and used graph associated matrices are the *adjacency matrix* and the *Laplacian matrix*. Two nodes v_k and v_l are adjacent to another if there exist an edge $e = \{v_k, v_l\}$. The entries $a_{k,l}$ of the adjacency matrix $A(G) = [a_{k,l}]_{n \times n}$ are 1, if there exists an edge $e = \{v_k, v_l\}$ (otherwise 0). The Laplacian matrix is defined as $L(G) = D(G) - A(G)$, where $D(G)[d_{k,l}]_{n \times n}$ is the degree matrix with $d_{k,k} = d(v_k)$ and $d_{k,l} = 0$, for all k unequal l.

The *weighted adjacency matrix* $A^\star(G) = [a^\star_{k,l}]_{n \times n}$ of a weighted graph G contains the labels associated with the edges, instead of just 1: $a^\star_{k,l} = w(\{v_k, v_l\})$ if $e = \{v_k, v_l\}$ in $E(G)$ (otherwise 0). Analogously, the *weighted Laplacian matrix* is defined as $L^\star(G) = D^\star(G) - A^\star(G)$, using the diagonal weighted degree matrix $D^\star(G)[d^\star_{k,l}]_{n \times n}$ with $d^\star_{k,k} = \sum_l a^\star_{k,l}$ (sum of all edge labels connected with the node v_k).

The *complement graph* $\overline{G} = (V, \overline{E}, \overline{w})$ of a graph $G = (V, E, w)$ incorporates the same node set V as G, and if two nodes are not adjacent in G, they are adjacent in $\overline{G}$. The *weighted adjacency matrix of the complement graph* will be named $\overline{A}(G)$ in the following.

3   Classification in Vector Spaces

There exist a huge amount of classification approaches to classify patterns in vector spaces [14],[15]. One of the most popular methods is the classification with support vector machines (SVMs) [16]. These supervised machine learning methods can be used for binary classification problems. During the training process, the SVM uses the given data points (training set) with the corresponding class labels (teachers) to optimize a hyperplane h, which separates the data points into the two given classes, whereby the two margins/gaps between the hyperplane and the data points should be as wide as possible. New data points are then mapped in the same space and the side of the hyperplane they belong to, decides their class labels.

We also applied a fuzzy output version of the SVMs. Therefore, the distance of a data point to the hyperplane is calculated (instead of just the algebraic sign) and normalized with a logistic sigmoid function. Hence, the output of the SVM can be treated as a probability to belong to the first class.

If there are more than two classes to distinguish, as is shown in Figure 1, it is not possible to use just one hyperplane to separate the classes, e.g. a three-class problem: *circles*, *stars* and *squares*.

There are two possibilities to solve this problem:

1. **One-vs.-One Classifier:**

   all pairs of possible two-class SVMs (1-2, 1-3, 2-3) are trained and the decision is made by voting.

2. **One-vs.-All Classifier:**

   every class is trained against all others (1-23, 2-13, 3-12) and the highest probability leads to the most likely class. Of course, this method just works for the fuzzy output SVMs.

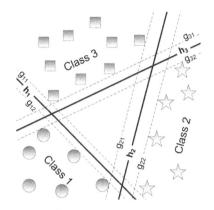

Fig. 1. Three-class problem, solved by a one-vs.-all classifier. The hyperplanes h_1, h_2 and h_3 separate the three classes.

Most often, the two classes are not linearly separable in their dimensional space. By applying a adequate function on the data points, they are mapped in a much higher dimensional space, where they are now linearly separable. The optimized linear hyperplane in this higher dimensional space is then transformed back in the lower dimensional space, where it becomes a non-linear hyperplane, but separates the classes. Because this method needs a lot of computation time, the so-called *kernel trick* is used. There are special functions, called *kernel functions*, which allow the optimization of the hyperplane without really transforming the data points. For more information on SVMs and RBFs see [16].

4   Data Description and Feature Extraction

The capital letter data set of the *IAM Graph Database Repository* [17] was utilized as application data set. It consists of graphs, representing capital letter drawings of all 15 "straight line" letters in the Roman alphabet (*A*, *E*, *F*, ...). In order to enforce differences in the samples, weak distortion operations, *shifting*, *removing* or *adding* lines to a prototype line drawing, were conducted. The nodes of the graphs represent the ending points of the lines in the drawing and the edges stand for the lines themselves. Finally, the nodes are labeled with their two-dimensional coordinates. The data set contains 2250 samples (150 for each class), equally divided in a test set, a validation set and a training set. Figure 2 shows examples of the graphical representation of the letter *A* with different kinds of distortion.

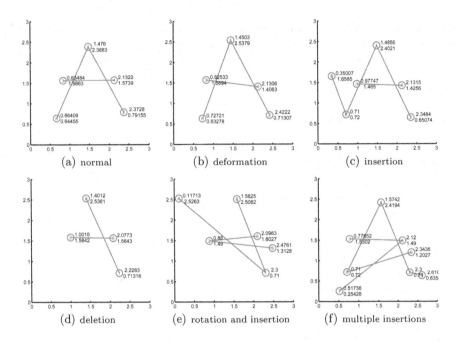

Fig. 2. Sample of the letter A with different kinds of distortion

In order to attain weighted graphs from the data, the following *labeling function* was used: $w : E(G) \rightarrow \mathbb{R}_0^+$. It attaches to each edge $e = \{v_k, v_l\}$ the Euclidean distance of the two nodes v_k and v_l.

Afterward, different matrices were extracted from the graph and their eigenvalues were computed. The eigenvalues $\{\lambda_1, ..., \lambda_p\}$ of the adjacency matrix $A(G)$ of a graph G build the so-called spectrum $SP(G) = [\lambda_1, ..., \lambda_p]$ and represent the zeros of the characteristic polynomial $|\lambda I - A(G)|$. For the eigenvalues and the corresponding eigenvectors $\{x_1, ...x_p\}$, the following equation (1) holds:

$$A(G)x_i = \lambda_i x_i, \, \forall \, i = 1, ..., p.\qquad(1)$$

Within one spectrum the eigenvalues were sorted in ascending order. To handle the different orders of the graphs, and therefore the different sizes of the spectra, we filled the too short spectra with zero values. The following spectra were computed:

SP_A Spectrum of the adjacency matrix
SP_L Spectrum of the Laplacian matrix
SP_{A^*} Spectrum of the weighted adjacency matrix
SP_{L^*} Spectrum of the weighted Laplacian matrix
$SP_{\overline{A^*}}$ Spectrum of the weighted adjacency matrix of $\overline{G}$
SP_{F^*} Spectrum of the filled weighted adjacency matrix (in this case, all distances were included, even if there were no edges, i.e. distance matrix of the nodes).

5   Experimental Results

In this section, the classification results for the six different feature types (as defined in Sect. 4) are presented and the performances are compared. We conducted experiments with One-vs.-One SVMs as well as with One-vs.-All SVMs. As kernel functions *RBF kernels*, *linear kernels*, *quadratic kernels* and *polynomial kernels* were investigated. The validation set allows us to find a suitable standard deviation σ for the RBF functions, which provided the best classification performance.

The decisions of the SVMs have to be calculated in different ways: For the hard results (class labels) we conducted a decision voting of the 105 One-vs.-One SVMs (One-vs.-One - hard). Because each SVM emit a probability for the first class (see Sect. 3), it is possible to compute the average rate for each class too (One-vs.-One - fuzzy). The fuzzy decision for the One-vs.-All SVMs can be obtained by choosing the maximum probability of the single classes. Table 1 shows the classification rates for the different feature types and the different SVM models with RBF kernel functions (exemplary with $\sigma = 0.8$).

Table 1. Classification results for the spectra SP_A, SP_L, SP_{A^*}, SP_{L^*}, $SP_{\overline{A^*}}$ and SP_{F^*}. For all SVMs, radial basis function kernels were used with a specific standard derivations $\sigma = 0.8$. Decisions for the classes by majority vote (One-vs.-One - hard), by computing the average classification rate for the classes (One-vs.-One - fuzzy) and by selecting the class with the maximum probability (One-vs.-All - fuzzy).

Feature type	One-vs.-One		One-vs.-All
	hard	fuzzy	fuzzy
SP_A	0.491	0.557	0.179
SP_L	0.232	0.475	0.192
SP_{A^*}	0.645	0.867	**0.897**
SP_{L^*}	0.493	0.803	0.799
$SP_{\overline{A^*}}$	0.292	0.825	0.845
SP_{F^*}	0.364	0.745	0.760

The hard decisions of the One-vs.-One SVMs lead to the lowest classification rates. It is unlikely that all 14 SVMs, which are responsible for one letter, emit this letter but it frequently happens that a few wrong decisions lead to a wrong winner. The emitted probabilities of the SVMs revealed, that the correct letter is often the second or the third winner. By including the probabilities in the calculation (fuzzy decisions) we could keep this information. The fuzzy decisions leads to a significant ($p = 0.03$) improvement of the classification results. Although, the One-vs.-All SVMs achieved the highest classification fuzzy rate (89.7 % for

$SP_{A^\star}$), the average fuzzy rate (for all spectra) of the One-vs.-One classifiers with 71.2 % is higher than the average of the One-vs.-All classifiers with 61.2 %.

Table 2 shows the best models for the six feature types, including the most suitable assignment for the standard deviation σ of the RBFs (range of σ between 0.5 and 3.0).

Table 2. Highest achieved classification rates for the different spectra SP_A, SP_L, $SP_{A^\star}$, $SP_{L^\star}$, $SP_{\overline{A^\star}}$ and $SP_{F^\star}$, the responsible model and the most suitable assignment for σ

Feature type	Classification method	σ	Classification rate
SP_A	One-vs.-One - fuzzy	0.8	0.557
SP_L	One-vs.-One - fuzzy	0.8	0.474
$SP_{A^\star}$	One-vs.-All - fuzzy	1.0	**0.899**
$SP_{L^\star}$	One-vs.-One - fuzzy	0.6	0.813
$SP_{\overline{A^\star}}$	One-vs.-All - fuzzy	1.0	0.847
$SP_{F^\star}$	One-vs.-One - fuzzy	1.5	0.777

The spectra of the adjacency matrices (SP_A) and the Laplacian matrices (SP_L), with no further information about the shape of the graph, provide the lowest classification rates. This is clear if one imagines the following simple example: the letter M and the letter W have the same number of nodes and the same edges between them. W is just a flipped version of M. Because the sorted eigenvalues and therefore the spectrum of these two matrices are invariant with respect to the labeling of the nodes, they are exactly the same.

By including all distances ($SP_{F^\star}$), the information about the characteristic shape of the letter gets lost. Letters with the same number of nodes only differ in the pairwise distances of the nodes but not in the number of non-zero entries in the matrices.

With the spectra of the Laplacian matrix ($SP_{L^\star}$) and the spectra of the adjacency matrix of the complement graph ($SP_{\overline{A^\star}}$), classification rates over 80 % could be achieved, but the spectra of the weighted adjacency matrices ($SP_{A^\star}$) provided the highest classification rate with 89.9 % (see Table 2). The information about the characteristic shape of the letters is included in all these features, i.e. where the edges are and also the lengths of the edges.

The difference between the recognition rates of the adjacency matrices (SP_A and $SP_{A^\star}$) and the Laplacian matrices (SP_L and $SP_{L^\star}$) is not significant ($p = 0.23$), however within this experiment, the matrices with the zeros on the diagonal A and $A^\star$ lead to a slightly higher performance.

Table 3 shows the confusion matrix of the 15 One-vs.-All SVMs for the spectrum of the weighted adjacency matrix $SP_{A^\star}$. The columns refer to the actual

Table 3. Confusion matrix of the 15 One-vs.-All SVMs for the feature SP_{A^*} (spectrum of the weighted adjacency matrix). The columns refer to the actual classes and the rows to the identified classes.

		A	E	F	H	I	K	L	M	N	T	V	W	X	Y	Z
									actual class							
identified class	A	46	3	0	0	0	0	2	0	0	0	0	0	0	0	0
	E	4	43	3	2	0	0	0	0	0	0	0	0	0	0	0
	F	0	2	47	1	0	4	0	0	0	0	0	0	0	0	0
	H	0	2	0	45	0	2	0	0	0	0	0	0	0	0	0
	I	0	0	0	0	41	0	3	0	0	0	0	0	0	0	0
	K	0	0	0	2	0	42	0	0	0	0	0	0	0	0	0
	L	0	0	0	0	8	1	43	1	0	0	0	0	0	0	0
	M	0	0	0	0	0	1	2	47	0	0	0	3	1	0	0
	N	0	0	0	0	0	0	0	0	47	0	2	1	0	0	6
	T	0	0	0	0	0	0	0	0	1	49	0	0	0	3	0
	V	0	0	0	0	0	0	0	0	1	0	46	1	1	0	0
	W	0	0	0	0	0	0	0	0	1	1	0	42	2	0	1
	X	0	0	0	0	0	0	0	0	0	0	0	0	46	0	0
	Y	0	0	0	0	0	0	1	2	0	0	0	3	0	47	0
	Z	0	0	0	0	1	0	1	0	0	0	2	0	0	0	43
		0.92	0.86	0.94	0.90	0.82	0.84	0.86	0.94	0.94	0.98	0.92	0.84	0.92	0.94	0.86

classes and the rows to the identified classes. The entries of the matrix are the number of letters (each letter appears 50 times in the data set). The last row contains the classification rates for each class.

The confusion matrix shows, that there are some letters, i.e. pairs of letters, which seem more difficult to distinguish than others. Almost 15 % (11 out of 76) of the misclassified letters are confusions of the letters *I-L*. The other eye-catching pairs are *A-E* with almost 10 %, *N-Z* with almost 8 % and *E-F* with more than 6 % of the misclassification rate. These results are not really surprising because the pair building letters are similar to each other, e.g. one edge more or rotated.

The experiments show, it is essential to choose an underlying matrix, which contains the important information to maintain the differences between the classes. In some cases, the crucial information may be the number of edges, then the adjacency matrix can be appropriate. In this case, the graphs vary in their shape, which depends on the existence of the edges and even more on their length and hence, it was necessary to include this relevant information in the features.

5.1 Decision Fusion of the Different Outputs

To probably enhance the classification performance, we computed combinations of the One.-vs.-All - fuzzy classification outputs of the different feature types. In some cases, the highest probability leads to a wrong class, but the second highest would have been the correct class. Therefore, we added the probabilities of the different SVMs class-wise, before we chose the maximum to get the identified class. The best three combinations with their overall classification are (for the individual classification performances, see Table 1):

1. $SP_{A^\star} + SP_{\overline{A^\star}} + SP_{F^\star}$ 92.4 %
2. $SP_{A^\star} + SP_{L^\star} + SP_{\overline{A^\star}} + SP_{F^\star}$ 92.1 %
3. $SP_A + SP_{A^\star} + SP_{L^\star} + SP_{\overline{A^\star}} + SP_{F^\star}$ 92.0 %

The results show that through combination of the individual classifiers on decision level, the misclassification rate can be decreased by 24 %. The combination with the SVMs, trained with $SP_{A^\star}$, $SP_{\overline{A^\star}}$ and $SP_{F^\star}$ has the highest overall classification rate.

5.2 Comparison with Related Work

To compare our results with other approaches, we selected papers where the same data set was utilized. In [18], bipartite graph matching for computing the edit distance was used and the distance measure was then utilized to classify the graphs with a 5-nearest-neighbor classifier. The highest classification rate with 84 % was achieved with their own method (Munkres' Algorithm). As reference method they implemented the optimal tree-search algorithm for computing the graph edit distance, introduced in [19] and [5], which achieved a recognition rate of 82.7 %. However, these are completely different methods, since these methods are graph-based and not feature-based.

As mentioned in the introduction, the actually best performance with 98.9 % on this data was achieved by [11]. In this paper, they "aim at bridging the gap between the domain of feature based and graph based object representations". They calculate the graph edit distances between one graph and m chosen prototypes and utilize this m-dimensional dissimilarity measure to classify the graph in a vector space using a SVM classifier. Although this method achieved a high classification performance, it is depending on a lot of variables and time-consuming calculations: the calculation of the graph edit distance is strongly connected to the order of the graphs (which makes it inappropriate for large graphs) and a good selection of the prototypes may be crucial for the classification process.

6   Discussion and Future Work

Using the spectrum of a graph as feature for classification is a well known method. The eigenvalues of graph-associated matrices can be utilized to embed the graphs in a vector space and use standardized methods, e.g. SVMs, for the classification. There are many matrices which can be extracted from of a graph and we investigated the power of six of them for classification: spectrum of the adjacency matrix (SP_A), spectrum of the Laplacian matrix (SP_L), spectrum of the weighted adjacency matrix (SP_{A^*}), spectrum of the weighted Laplacian matrix (SP_{L^*}), spectrum of the weighted adjacency matrix of $\overline{G}$ ($SP_{\overline{A^*}}$) and the spectrum of the filled weighted adjacency matrix (SP_{F^*}). As label for each edge, the Euclidean distances between the two connected nodes were utilized.

As data set, we used the capital letter data set of the *IAM Graph Database Repository* (see Sect. 4) because is has become a benchmark data set for graph classification and our results can easily be compared to other approaches.

The weighted versions (SP_{A^*},SP_{L^*} and also $SP_{\overline{A^*}}$) include more information than the others and generate classification rates between 80 % and 90 %. The misclassified letters build confusion pairs, that means, there are some couples which are difficult to distinguish. The following pairs *I-L*, *A-E*, *N-Z* and *E-F* cause almost 40 % of the misclassified letters. One reason for the poor distinctness is, that the letters are similar to each other (e.g. just one edge more) and the other reason is that the spectrum is invariant with respect to the rotation of the graph, more precisely the labeling of the nodes (*N* ist just a rotated *Z*). In consideration of this fact, the classification performance is absolutely satisfying.

By combining the outputs depending on different features (decision fusion), the misclassification rate can be decreased by 24 %. Each of the underlying matrices contain information, which the others lack and the combination can benefit from all. With a classification rate of 92.4 % we are below the best classification performance on the data set with 98.9 % [11]. However, we did not aim at outperforming other approaches, but to investigate the power of different graph-associated matrices for classification. It is absolutely important to choose the graph-associated matrix which contains the diverse information needed for the classification task and in this case, the length between the nodes is highly relevant.

The next step is to utilize more complex data sets and investigate the eigenvalues of the weighted adjacency matrices (weighted with the Euclidean distances) in a more theoretical way. Is there a connection between a eventually existing distribution of the distances in the matrices and the eigenvalues of them? How big is the influence of the connectivity in the graph? We started our research with an artificial generated data set.

Acknowledgments. The work of Miriam Schmidt is supported by a scholarship of the Graduate School *Mathematical Analysis of Evolution, Information and Complexity* of the University of Ulm. The work of Günther Palm and Friedhelm Schwenker is supported by the Transregional Collaborative Research Center *SFB/ TRR 62 Companion-Technology for Cognitive Technical Systems*, funded by the German Research Foundation (DFG).

References

1. Cvetković, D.M., Doob, M., Sachs, H.: Spectra of Graphs: Theory and Applications, 3rd edn. Vch Verlagsgesellschaft Mbh (1998)
2. Chung, F.R.K.: Spectral Graph Theory. CBMS Regional Conference Series in Mathematics, vol. 92. Oxford University Press (1997)
3. Brouwer, A.E., Haermers, W.H.: Spectra of Graphs. Universitext. Springer (2012)
4. Garey, M.R., Johnson, D.S.: Computers and Intractability; A Guide to the Theory of NP-Completeness. W. H. Freeman & Co. (1990)
5. Sanfeliu, A., Fu, K.S.: A distance measure between attributed relational graphs for pattern recognition. IEEE Transactions on Systems, Man, and Cybernetics 13(3), 353–362 (1983)
6. Bunke, H., Shearer, K.: A graph distance metric based on the maximal common subgraph. Pattern Recognition Letters 19(3-4), 255–259 (1998)
7. Luo, B., Wilson, R.C., Hancock, E.R.: Spectral embedding of graphs. Pattern Recognition 36(10), 2213–2230 (2003)
8. Wilson, R.C., Hancock, E.R., Luo, B.: Pattern vectors from algebraic graph theory. IEEE Trans. Pattern Anal. Mach. Intell. 27(7), 1112–1124 (2005)
9. Schmidt, M., Schwenker, F.: Classification of Graph Sequences Utilizing the Eigenvalues of the Distance Matrices and Hidden Markov Models. In: Jiang, X., Ferrer, M., Torsello, A. (eds.) GbRPR 2011. LNCS, vol. 6658, pp. 325–334. Springer, Heidelberg (2011)
10. Umeyama, S.: An eigendecomposition approach to weighted graph matching problems. IEEE Transactions on Pattern Analysis and Machine Intelligence 10(5), 695–703 (1988)
11. Riesen, K., Neuhaus, M., Bunke, H.: Graph Embedding in Vector Spaces by Means of Prototype Selection. In: Escolano, F., Vento, M. (eds.) GbRPR 2007. LNCS, vol. 4538, pp. 383–393. Springer, Heidelberg (2007)
12. Diestel, R.: Graph Theory, 3rd edn. Graduate Texts in Mathematics, vol. 173. Springer (2005)
13. Bollobás, B.: Modern Graph Theory, 2nd edn. Graduate Texts in Mathematics. Springer (2002)
14. Vapnik, V.N.: The nature of statistical learning theory, 2nd edn. Statistics for Engineering and Information Science. Springer (1999)
15. Bishop, C.M.: Pattern Recognition and Machine Learning (Information Science and Statistics). Springer (2007)
16. Scholkopf, B., Smola, A.J.: Learning with Kernels: Support Vector Machines, Regularization, Optimization, and Beyond. Adaptive Computation and Machine Learning. MIT Press (2002)
17. Riesen, K., Bunke, H.: IAM Graph Database Repository for Graph Based Pattern Recognition and Machine Learning. In: da Vitoria Lobo, N., Kasparis, T., Roli, F., Kwok, J.T., Georgiopoulos, M., Anagnostopoulos, G.C., Loog, M. (eds.) SSPR&SPR 2008. LNCS, vol. 5342, pp. 287–297. Springer, Heidelberg (2008)
18. Riesen, K., Neuhaus, M., Bunke, H.: Bipartite Graph Matching for Computing the Edit Distance of Graphs. In: Escolano, F., Vento, M. (eds.) GbRPR 2007. LNCS, vol. 4538, pp. 1–12. Springer, Heidelberg (2007)
19. Bunke, H., Allermann, G.: Inexact graph matching for structural pattern recognition. Pattern Recognition Letters 1(4), 245–253 (1983)

Improving Iris Recognition through New Target Vectors in MLP Artificial Neural Networks

José Ricardo Gonçalves Manzan, Shigueo Nomura, Keiji Yamanaka,
Milena Bueno Pereira Carneiro, and Antônio C. Paschoarelli Veiga

Faculty of Electrical Engineering - Federal University of Uberlândia,
Av. João Naves de Ávila, 2160, Bloco 3N - Campus Santa Mônica
CEP: 38400-902 - Uberlândia - MG, Brasil
josericardo@iftm.edu.br, shigueonomura@feelt.ufu.br,
{keiji,acpveiga}@ufu.br, milenabueno@yahoo.com.br
http://www.feelt.ufu.br

Abstract. This paper compares the performance of multilayer perceptron (MLP) networks trained with conventional bipolar target vectors
(CBVs) and orthogonal bipolar new target vectors (OBVs) for biometric
pattern recognition. The experimental analysis consisted of using biometric patterns from CASIA Iris Image Database developed by Chinese
Academy of Sciences - Institute of Automation. The experiments were
performed in order to obtain the best recognition rates, leading to the
comparison of results from both conventional and new target vectors.
The experimental results have shown that MLPs trained with OBVs
can better recognize the patterns of iris images than MLPs trained with
CBVs.

Keywords: Biometric pattern, iris image, conventional bipolar vector,
multilayer perceptron, orthogonal bipolar vector, pattern recognition,
target vector.

1 Introduction

Among the various research fields in computing, computational intelligence has
received enough attention on the possibilities of applications to genetic algorithms, fuzzy logic, and artificial neural networks (ANNs).

Research advances in ANNs have been realized from the 80's mainly due to
the relevant contributions of expert researchers specialized in this field. However,
it is evident that the researchers still are searching for further advances through
several studies [1–5].

Several studies in pattern recognition techniques such as statistical approach
[6] and connectionist approach [6] have been performed. The connectionist approach has involved ANN techniques providing relevant and promising results
in pattern recognition tasks due to their generalization capabilities.

Among the ANN techniques, the MLP models have been widely applied to
biometric pattern recognition problems. MLPs have been focused by researchers
of this field around the world.

N. Mana, F. Schwenker, and E. Trentin (Eds.): ANNPR 2012, LNAI 7477, pp. 115–126, 2012.
© Springer-Verlag Berlin Heidelberg 2012

Some studies have been related to MLP topology [2] choice improvements. Also, studies for investigating improvements in the learning algorithm [3] are not rare. Furthermore, interesting results have been published due to a new methodology for choosing the initial synaptic weights [4] and use of signal sensitivity analysis [1].

Searching for mechanisms that can improve the performance of MLPs in biometric pattern recognition tasks, we could not find investigations focused on target vector studies.

Usually, we have the conventional use of bipolar vectors (CBVs) as targets for MLP training and one-per-class classification approach.

This paper proposes an unconventional use of orthogonal bipolar vector (OBV) as targets for training MLP models. We suppose that the use of new target vectors can influence on MLP performance improvement in recognizing biometric patterns. Mathematically, OBVs have advantages of reduced similarities between them.

Preliminary experimental results related to this proposal have been presented by previous works [7–10] and they have shown the effectiveness for the MLP performance improvement. This paper aims to show experimental results of this new approach applied to iris image recognition.

1.1 Motivation

It is known that biological neurons can recognize patterns with a high degree of degradation [11]. This ability to differentiate degraded patterns can also be obtained by MLPs trained with appropriate and adjusted conditions. An MLP model can recognize a degraded or biometric pattern even though it has not been presented during the training stage. Therefore, the biometric pattern recognition by MLPs is possible if the models provide a good generalization capability.

There are several proposals searching for the appropriate treatment of input vectors [12] to achieve the expected MLP performance improvement. However, there is no relevant investigation focusing on the treatment of target vectors for MLP learning. Regarding this investigation gap, we have decided to investigate the positive influence of OBVs used as targets on the MLP performance to recognize biometric patterns.

1.2 Mathematical Foundation

Analyzing conventional target vectors in terms of inner product between them, we can verify that the product increases according to the size of those vectors. Also, the inner product is high if the angle between two consecutive vectors is small. In other words, if the vectors are almost parallel then their degree of similarity is high.

On the other hand, if the angle between two consecutive vectors is the same as 90 degrees then the inner product between them is null. In this case, the vectors are orthogonal and the Euclidean distance between them is large compared to the conventional vectors.

In terms of similarity, we can realize that OBVs have null similarity, regardless of their sizes. Eq. 1 and eq. 2 represent two possible target vectors and eq. 3 represents the corresponding inner product between them. The Euclidean distance can be calculated by eq. 4.

$$\vec{V_i} = (v_1, v_2, \ldots, v_n) \tag{1}$$

$$\vec{W_i} = (w_1, w_2, \ldots, w_n) \tag{2}$$

$$\vec{V_i} \bullet \vec{W_i} = v_1 \cdot w_1 + v_2 \cdot w_2 + v_3 \cdot w_3 + \ldots + v_n \cdot w_n \tag{3}$$

$$d_{V,W} = \sqrt{(w_1 - v_1)^2 + (w_2 - v_2)^2 + (w_3 - v_3)^2 + \ldots + (w_n - v_n)^2} \tag{4}$$

In terms of Euclidean distance, it is easy to verify that the distance between two CBVs is constant for any size of vectors. But, the distance between two OBVs increases when their sizes increase. Also, the Euclidean distance between OBVs is always larger than the distance between CBVs.

Section 2 presents the experimental procedure. The experimental results are presented in Section 3. In Sections 4 and 5, the discussion of results and conclusion are described.

2   Experimental Procedure

This section presents the experimental procedure for MLP training and iris image recognition using three model types. The first type of model is the MLP trained with CBVs, the second type is the model trained with OBVs, and the third one is the model trained with NOVs. The experiments aim to do the hypothesis confirmation for the performance improvement of MLPs trained with OBVs as targets to recognize iris images as biometric patterns.

2.1   Target Vectors for MLP Learning

A set of conventional bipolar vectors (CBVs) can be defined as a matrix described by eq. 5. Each row i of this matrix corresponds to the i-th CBV containing the component "1" for $i = j$ and the component "-1" for others.

$$\vec{V_{ij}} = \begin{cases} 1 & \text{for } i = j \\ -1 & \text{for } i \neq j \end{cases} \tag{5}$$

To generate a set of orthogonal bipolar vectors (OBVs), we have based on the theorem and its respective algorithm presented by Fausett [13]. According to the theorem, the number n of components for a vector is calculated as $n = 2^k m$. The value of m is the number of components of seed vector V that starts the algorithm. The value of 2^k is the number of generated orthogonal vectors. The operation $[V, V]$ denotes the function for concatenation of vector V with itself producing a vector with double number of components. The steps of the algorithm can be described as follows:

1. Initializing the seed vector: $V_m(1) = (1, 1, 1, \ldots, 1)$ where m is odd;
2. Concatenating the seed vectors and generating mutually orthogonal vectors:
   $V_{2m}(1) = [V_m(1), V_m(1)]$ and $V_{2m}(2) = [V_m(1), -V_m(1)]$;
3. Concatenating consecutive orthogonal vectors:
   $V_{4m}(1) = [V_{2m}(1), V_{2m}(1)]$, $V_{4m}(2) = [V_{2m}(1), -V_{2m}(1)]$,
   $V_{4m}(3) = [V_{2m}(2), V_{2m}(2)]$, $V_{4m}(4) = [V_{2m}(2), -V_{2m}(2)]$;
4. Repeating step 3 until generating $V_n(1), \ldots, V_n(2^k)$ as OBVs.

A sample of generated OBVs with eight components can be as follows:

$V_{8m}(1) = (1, \ 1, \ 1, \ 1, \ 1, 1, \ 1, 1)$, $V_{8m}(2) = (1, \ 1, \ 1, \ 1, -1, -1, -1, -1)$,
$V_{8m}(3) = (1, \ 1, -1, -1, \ 1, \ 1, -1, -1)$, $V_{8m}(4) = (1, \ 1, -1, -1, -1, -1, \ 1, \ 1)$,
$V_{8m}(5) = (1, -1, \ 1, -1, \ 1, -1, \ 1, -1)$, $V_{8m}(6) = (1, -1, \ 1, -1, -1, \ 1, -1, \ 1)$,
$V_{8m}(7) = (1, -1, -1, \ 1, \ 1, -1, -1, \ 1)$, $V_{8m}(8) = (1, -1, -1, \ 1, -1, \ 1, \ 1, -1)$.

In this experimental analysis, we also use the non-orthogonal bipolar vectors (NOVs) that have the same size as the OBVs, but they have non-null inner product. The difference of NOVs in relation to CBVs is only the size. The use of NOVs is a strategy for fair performance comparison with larger sizes of OBVs in relation to CBVs.

If the size of NOVs is large then the corresponding inner product between them is large. Since the inner product represents the degree of similarity between NOVs, the similarity increases while their sizes increase. We are supposing that the similarity between target vectors can cause low performance of MLPs.

2.2 MLP Topologies

Cross validation methods were applied to define the MLP topologies for the experiments. So, four MLP topologies (number of input neurons x number of hidden neurons x number of output neurons) were set as follows: 2400 x 200 x 71; 2400 x 200 x 128; 2400 x 800 x 71; and 2400 x 800 x 128.

The number of 71 units in the output layer was defined by the number of subjects corresponding to the iris images from CASIA repository. So, 71 CBVs with 71 components were generated to represent the target vectors for the usual pattern recognition experiments.

In case of using OBVs as target vectors, since OBVs have been generated with the sizes same as 2, 4, 8, 16, 32, 64, 128 and so on, we have chosen the number of 128 units in the output layer. Therefore, we have generated OBVs with 128 components for the experiments regarding the proposed approach. Based on the size of OBV, we have generated NOVs with 128 components for the third type of MLP model.

2.3   Experimental Data

The training data consisted of human iris images from the Chinese Academy of Sciences - Institute of Automation database called CASIA [14]. The database contains iris images from 108 subjects and 71 of them consist of complete data with seven images. For this reason, we have adopted the data corresponding to 71 subjects. Randomly chosen four images composed the training set and other three ones composed the test set. According to CASIA repository, these images were taken by the use of infrared light to get the iris features with enough contrast for biometric pattern recognition.

The iris image processing takes place in a few steps. The first step is the location of the iris' region in the image that is performed using the circular Hough transformed. Then the iris' region, which has a ring shape, is normalized to be represented by a rectangular matrix. Finally, the extraction of the iris' features is done. In this paper, the iris' features were extracted convoluting the normalized image with the so-called log Gabor filter. Filtration gives rise to complex coefficients whose phases are quantized to one of the four quadrants of the complex plane. Each quadrant is referenced by two bits, and a binary template is created [15–17]. For each image there are 8640 pixels arranged in 18 concentric circles each containing 480 pixels.

In this work, we extracted only 5 concentric circles of the iris, eliminating the interference of the eyelids and reducing the training effort of the MLP. Thus, each training pattern corresponds to a set of 5 x 480 = 2400 pixels. The white pixels were represented by value -1 and the black pixels by 1. So, the training vectors were constructed to represent single lines containing 2400 pixels that connect the points from the innermost circle to the outermost one of the iris.

2.4   MLP Training Stage

The experimental simulations were performed using the *traingdx* toolbox of Matlab software version R2008. The *traingdx* toolbox adopts the momentum and adaptive learning rate for MLP learning. For this reason, we can get a more rapid convergence assuring consistent results. As the toolbox uses randomly chosen initial synaptic weights, we have slightly different final weights for different simulations. So, we have performed three different simulations per each set of training parameters to assure a better representation of the experimental results shown in Tables 1 and 2.

The adopted initial learning rates were 0.1 and 0.3, since *traingdx* works with adaptive learning rates. A tolerance for error during the training epochs was adopted as stopping criterion.

The equations for calculating the mean squared error for using the stopping criterion are as follows:

- $E_p = \frac{1}{2} \sum_{j=1}^{Ns} (d_j - y_j)^2$ where E_p is the squared error of a pattern p; Ns is the number of output neurons; d_j is the desired output for neuron j; and y_j is the net output for neuron j;
- $E_m = \frac{1}{N} \sum_{p=1}^{N} E_p$ where E_m is the mean squared error of all patterns for each epoch; N is the number of patterns; E_p is the squared error for pattern p;

The training has been finalized when the stopping condition $E_m < \varepsilon$ is satisfied. The value of ε is the initialized tolerance for error during the training process.

The simulations were performed on a computer's processor type INTEL(R) CORE i5-2410TM, 2.30 GHz and memory (RAM) of 4 GB.

3   Experimental Results

Tables 1 and 2 show the results from the MLP training and test using the Matlab toolbox. Tolerance in the tables means the maximum mean squared error (represented by ε in Section 2.4) to be achieved by the MLP training for satisfying the stopping condition. Each table provides the tolerance for error, the number of epochs for the training, the pattern recognition rate (performance), the simulation number, and the initial learning rate α .

In Tables 1 and 2, we can verify that results from 3 simulations for a set of parameters are presented and these results present small standard deviation between them. The small difference between those results is due to the random initialization of synaptic weights for the MLP training with Matlab toolbox.

Table 1 compares the performance of three different types of target vectors with 200 hidden neurons and initial learning rate of 0.1 or 0.3. In Table 2, the results refer to the adoption of 800 hidden neurons. The bold and underlined value represents the best obtained recognition rate for a set of simulations.

Fig. 1 represents the best simulation results for each type of target vector setting up MLP with 200 hidden neurons and initial learning rate of 0.1. Fig. 2 corresponds to the results from simulations of MLPs with 200 hidden neurons and initial learning rate of 0.3. The best simulation results corresponding to experiments using MLPs with 800 hidden neurons and initial learning rate of 0.1 are presented in Fig. 3. Fig. 4 presents the best experimental results for each target vector type in MLP with 800 hidden neurons and initial learning rate of 0.3.

In another application [7], the recognition rate of characters from license plate degraded images with the use of OBVs as targets has increased around 5.4% comparing with the results using conventional target vectors. The work applied to the recognition of handwritten digits [9] has presented a variation on the recognition rate by 2% when OBVs are used as target vectors for MLP learning. The OBV experimental results for application to the digits extracted from license plate degraded images [10] have presented an increase of 8% on the MLP performance. All these results have strengthened the viability of using OBVs as target vectors for MLP in pattern recognition.

Table 1. Comparison of MLP performances with 200 hidden neurons for various target vectors and initial learning rates (represented by α)

			Tolerance	1.00E-02	1.00E-03	1.00E-04	1.00E-05	1.00E-06
$\alpha=0.1$	OBV(128)	Simulation 1	Epochs	121	335	567	408	560
			Performance (%)	93.90	94.37	94.37	95.31	94.37
		2	Epochs	125	203	227	707	370
			Performance (%)	93.90	92.96	**95.77**	93.90	94.84
		3	Epochs	126	177	201	337	482
			Performance (%)	95.31	93.90	93.90	94.37	94.37
	NOV(128)	1	Epochs	81	128	215	189	239
			Performance (%)	47.89	74.65	86.85	84.04	86.85
		2	Epochs	82	153	194	187	249
			Performance (%)	46.48	78.40	84.04	82.16	**88.26**
		3	Epochs	87	127	243	236	313
			Performance (%)	40.85	68,54	85.92	81.69	83.57
	CBV(71)	1	Epochs	82	146	171	184	248
			Performance (%)	56.34	74.65	83.57	82.16	86.39
		2	Epochs	82	124	173	190	235
			Performance (%)	55.67	70.89	84.98	81.69	**88.73**
		3	Epochs	81	137	138	183	258
			Performance (%)	56.81	70.42	76.06	86.39	86.39
$\alpha=0.3$	OBV(128)	1	Epochs	105	188	598	563	359
			Performance (%)	**95.77**	94.37	93.90	93.90	94.37
		2	Epochs	100	149	535	262	435
			Performance (%)	94.84	94.37	92.96	92.96	94.84
		3	Epochs	102	202	450	470	819
			Performance (%)	94.37	94.37	95.31	93.43	93.84
	NOV(128)	1	Epochs	74	148	144	213	219
			Performance (%)	46.48	79.34	73.71	86.85	**89.67**
		2	Epochs	74	148	144	213	219
			Performance (%)	47.89	78.40	86.39	84.98	86.85
		3	Epochs	59	125	159	343	234
			Performance (%)	46.01	75.59	79.34	88.26	89.20
	CBV(71)	1	Epochs	76	122	115	174	213
			Performance (%)	65.26	78.40	79.81	84.51	88.26
		2	Epochs	79	132	115	163	294
			Performance (%)	63.85	82.63	76.53	83.10	86.85
		3	Epochs	63	159	131	163	262
			Performance (%)	59.62	74.65	81.69	86.39	**89.20**

Table 2. Comparison of MLP performances with 800 hidden neurons for various target vectors and initial learning rates (represented by α)

			Tolerance	1.00E-02	1.00E-03	1.00E-04	1.00E-05	1.00E-06
$\alpha = 0.1$	OBV(128) Simulation	1	Epochs	95	136	179	225	272
			Performance (%)	94.37	**96.24**	95.31	95.77	95.31
		2	Epochs	98	133	220	226	271
			Performance (%)	94.37	93.90	**96.24**	95.31	95.31
		3	Epochs	99	141	203	225	271
			Performance (%)	93.43	95.31	95.31	95.31	95.77
	NOV(128) Simulation	1	Epochs	95	142	185	224	307
			Performance (%)	44.60	78.40	80.75	86.39	87.32
		2	Epochs	86	142	182	179	212
			Performance (%)	47.42	73.71	77.46	82.63	**88.26**
		3	Epochs	91	139	234	273	230
			Performance (%)	53.05	75,12	83.57	86.85	84.98
	CBV(71) Simulation	1	Epochs	141	153	189	241	202
			Performance (%)	66.20	84.04	83.10	90.14	**91.55**
		2	Epochs	140	130	162	168	231
			Performance (%)	65.73	77.46	86.85	84.98	86.85
		3	Epochs	140	116	159	159	208
			Performance (%)	65.73	76.53	84.04	84.51	85.45
$\alpha = 0.3$	OBV(128) Simulation	1	Epochs	76	116	157	202	249
			Performance (%)	94.37	94.84	94.84	95.31	95.31
		2	Epochs	77	111	157	235	265
			Performance (%)	94.37	94.84	94.84	96.24	95.77
		3	Epochs	75	122	157	203	249
			Performance (%)	94.37	94.37	94.37	93.90	**96.71**
	NOV(128) Simulation	1	Epochs	109	157	342	619	732
			Performance (%)	58.69	84.51	86.39	90.14	**91.55**
		2	Epochs	96	157	390	424	310
			Performance (%)	58.69	84.98	90.14	90.61	87.80
		3	Epochs	106	166	361	350	569
			Performance (%)	58.69	86.39	89.20	89.20	90.14
	CBV(71) Simulation	1	Epochs	49	147	121	163	275
			Performance (%)	47.42	69.01	70.42	76.53	**85.45**
		2	Epochs	45	89	100	170	261
			Performance (%)	42.72	68.08	78.40	77.47	84.51
		3	Epochs	44	82	95	152	213
			Performance (%)	43.66	68.54	70.89	82.63	84.51

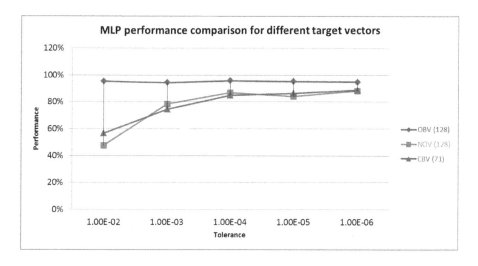

Fig. 1. 200 hidden neurons and initial learning rate of 0.1

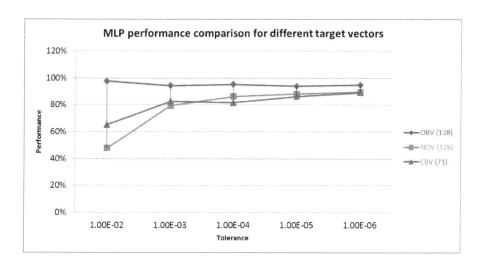

Fig. 2. 200 hidden neurons and initial learning rate of 0.3

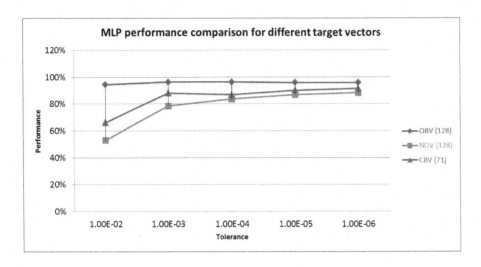

Fig. 3. 800 hidden neurons and initial learning rate of 0.1

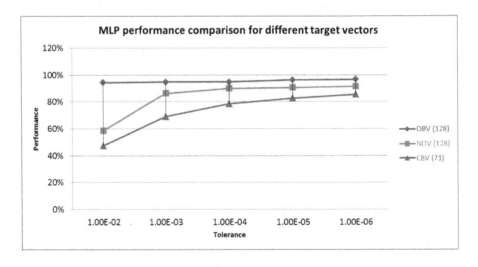

Fig. 4. 800 hidden neurons and initial learning rate of 0.3

4   Discussion

The results have shown that the use of OBVs as targets improves the performance on recognizing biometric patterns represented by iris images in all the cases.

We have verified that initial learning rates have caused some differences in MLP performance. Among others, the use of 800 neurons provided the best recognition rate which was 96.71% with initial learning rate of 0.3. Also, we have realized that recognition rates were higher than 93% using OBVs and rates from 47% to 66% using other vectors for MLP training in non-rigid tolerance (1.00E-02).

Furthermore, a large number of epochs were necessary to achieve the rates around 90% in case of using CBVs and NOVs as target vectors.

5   Conclusion

This paper proposed the unconventional use of OBVs as target vectors for MLP learning to improve the network performance on biometric pattern recognition. For this purpose, iris images from CASIA database were used to do MLP performance comparisons among CBVs, NOVs and OBVs as target vectors.

The experimental results showed high recognition rates using OBVs rather than other vectors. Also, the simulations showed that the use of OBVs provides smaller number of epochs for MLP learning compared to the use of other vectors. Consequently, the computational load can be reduced with the use of OBVs for MLP learning.

Another advantage of OBVs is related to the null inner product or null similarity between them while the inner product between conventional target vectors increases when the number of biometric patterns increases.

Therefore, we have concluded that the unconventional use of OBVs as target vectors for MLP learning and biometric pattern (represented by iris images) recognition is promising. The Euclidean distance increase and similarity reduction between vectors provided by the use of OBVs can support the obtained results.

Acknowledgements. We thank the PROPP in the Federal University of Uberlândia for supporting this work through the project number 72/2010. We also thank CAPES for supporting the Inter-institutional Master's program organized by the Federal University of Uberlândia and the Federal Institute of Triângulo Mineiro.

References

1. Wang, X., Chang, C., Du, F.: Achieving a More Robust Neural Network Model for Control of a MR Damper by Signal Sensitivity Analysis. Neu. Comp. & App. 13, 330–338 (2002)
2. Costa, M.A., Braga, A.P., Menezes, B.R.: Improving Neural Networks Generalization With New Constructive and Pruning Methods. J. Int. & Fuzzy Syst. 13, 75–83 (2003)
3. Lee, C.M., Yang, S.S., Ho, C.L.: Modified Back-propagation Algorithm Applied to Decision-feedback Equalization. IEE Proceedings - Vis., Ima. & Sig. Proc. 153, 805–809 (2006)

4. Kim, D.: Improving Prediction Performance of Neural Networks in Pattern Classification. Int. J. Comp. Math. 82(4), 391–399 (2005)
5. Chen, L., Pung, H.K.: Convergence Analysis of Convex Incremental Neural Networks. Annals of Mathematics and Artificial Intelligence 52, 67–80 (2008)
6. Browne, A.: Neural Network Analysis, Architetures, and Applications. Institute of Physics Pub., Philadelphia (1997)
7. Nomura, S., Yamanaka, K., Katai, O., Kawakami, H., Shiose, T.: Improving MLP Learning Via Orthogonal Bipolar Target Vectors. J. Adv. Comp. Int. and Int. Inf. 9, 580–589 (2005)
8. Nomura, S., Yamanaka, K., Katai, O., Kawakami, H., Shiose, T.: A New Approach to Improving Math Performance of Artificial Neural Networks (in Portuguese). In: VIII Brazilian Symposium on Neural Networks, São Luís (2004)
9. Manzan, J.R.G., Yamanaka, K., Nomura, S.: Improvement in Perfomance of MLP Using New Target Vectors (in Portuguese). In: X Brazilian Congress on Computational Intelligence, Fortaleza (2011)
10. Nomura, S., Manzan, J.R.G., Yamanaka, K.: An Experimentation With Improved Target Vectors for MLP in Classifying Degraded Patterns. Learning and Nonlinear Models 8(4), 240–252 (2010)
11. Cooper, L.N.: A Possible Organization of Animal Memory and Learning. In: Lundquist, B., Lundquist, S. (eds.) Nobel Symposium on Collective Propertiers of Physical Systems, Sweden, pp. 62–84 (1973)
12. Duda, R.O., Hart, P.E.: Pattern Classification and Scene Analysis. Wiley, New York (1973)
13. Fausset, L.: Fundamentals of Neural Networks: Architecture, Algorithms, and Applications. Prentice-Hall (1994)
14. Chinese Academy of Sciences - Institute of Automation, Database of 756 Greyscale Eye Images, http://www.cbsr.ia.ac.cn/IrisDatabase.html
15. Daugman, J.: High Confidence Visual Recognition of Person by a Test of Statistical Independence. IEEE Transactions on Pattern Analysis and Machine Intelligence 15(11), 1148–1161 (1993)
16. Negin, M., Chmielewski Jr., T.A., Salganicoff, M., von Seelen, U.M., Venetainer, P.L., Zhang, G.G.: An Iris Biometric System for Public and Personal Use. IEEE Computer Society 33, 70–75 (2000)
17. Carneiro, M.B.P., Veiga, A.C.P.: Application of Genetic Algorithms to Improve the Realiability of An Iris Recognition System. In: IEEE Workshop on Machine Learning for Signal Processing, Mystic, pp. 159–164 (2005)

Robustness of a CAD System on Digitized Mammograms

Antonio García-Manso[1], Carlos J. García-Orellana[1],
Ramón Gallardo-Caballero[1], Nico Lanconelli[2], Horacio González-Velasco[1],
and Miguel Macías-Macías[1]

[1] Pattern Classification and Image Analysis Group (CAPI),
University of Extremadura,
Avenida de Elvas, s/n., Badajoz, Extremadura, Spain
antonio@capi.unex.es

[2] Medical Imaging Group, Physics Dpt. - Bologna University, Viale B., Pichat 6/2,
40127 Bologna, Italy

Abstract. In this paper we study the robustness of our CAD system, since this is one of the main factors that determine its quality. A CAD system must guarantee consistent performance over time and in various clinical situations. Our CAD system is based on the extraction of features from the mammographic image by means of Independent Component Analysis, and machine learning classifiers, such as Neural Networks and Support Vector Machine. To measure the robustness of our CAD system we have used the digitized mammograms of the USF's DDSM database, because this database was built by digitizing mammograms from four different institutions (four different scanner) during more than 10 years. Thus, we can use the mammograms digitized with one scanner to train the system and the remaining to evaluate the performance, what gives us a measure of the robustness of our CAD system.

Keywords: ICA, NN, SVM, CAD.

1 Introduction and Purpose

The goal of a CAD system is to be able to analyze an image and indicate the location of possible lesions, if any. But also, there are several factors that should provide a CAD system to detect and diagnose masses in mammograms, such as: high sensitivity to detect the largest possible number of cancers, high specificity to reduce the number of false positives per image, acceptable call rate, early detection to increase the chances of survival, low processing time and robustness [1]. That is, the system must guarantee consistent performance along time and in various clinical situations. We have designed and implemented a CAD system to detect and classify masses in mammograms. We have used the USF's DDSM database [2] to train and test our CAD system.

The DDSM contains mammograms obtained from examinations between October 1988 and February 1999 in four different clinical sites: Massachusetts General Hospital (MGH) in Boston, Wake Forest University School of Medicine

N. Mana, F. Schwenker, and E. Trentin (Eds.): ANNPR 2012, LNAI 7477, pp. 127–138, 2012.
© Springer-Verlag Berlin Heidelberg 2012

(WFU) in North Carolina, Sacred Heart Hospital in Pensacola (SH), Florida, and Washington University School of Medicine in St. Louis Medical Center (WU). These mammograms were digitized using four different scanners: DBA M2100 ImageClear with a resolution of $42\mu m$, Howtek with a resolution of $43.5\mu m$, Lumisys 200 laser with a resolution of $50\mu m$ and Howtek MultiRAD 850 with a resolution of $43.5\mu m$. With that we can get an idea of the heterogeneous that is the used dataset. But, to normalize and avoid this heterogeneity in the used dataset we use the calibration curves which are available of each scanner to obtain the mammographic images in optical densities. In that way, at least in theory, we will have all dataset normalized in the same conditions. But the number of prototypes regarding to each class digitized with each scanner can be very different considering a scanner, or other. And, also, the way in which were indicated the ground truth over the mammograms could be very different [3], because of the long period of time (more than 10 years) during which was built the DDSM database. That is, in the DDSM database we can see different styles when the radiologists indicate the lesions on mammograms. First, because, surely, many radiologists were involved, and in addition, because it was done considering informed mammograms of four different institutions.

We propose a system to detect masses in mammograms as a two-class pattern recognition problem (mass or normal tissue), but, in our proposed approach, no modelling has been used. In contrast, we have used features extraction based on Independent Component Analysis (ICA), for its ability to obtain a basis functions adapted to the problem, especially to the natural images [4,5]. Thus, we have obtained some basis functions (basis images) to expand the original image (original patch), where the coefficients of this expansion will be used to form the input vectors to the classifiers.

The rest of our paper is organized as follows. Section 2 introduces the general concepts of feature extractions, the classifiers and the dataset used in our experiments. Section 3 includes a description of our methodology. Section 4 describes our results. Finally, Section 5 presents the main conclusions of this work.

2   Methods

In this section, we provide a brief description of the mammogram database utilized. Additionally we describe the procedure implemented to build a set of mass prototypes and the main characteristics of the selected image feature extractor. Finally, we provide a short description of the used classifiers.

2.1   Data and Prototype Creation

The Digital Database for Screening Mammography [2] is a resource available to the mammographic image analysis research community. Contains a total of 2,620 cases. Each case provides four screening views, mediolateral oblique (MLO) and craniocaudal (CC) projections of left and right breasts. Therefore, the database has a total of 10,480 images.

Cases are categorized in four major groups: *normal, cancer, benign and benign without callback*.

All cases in the DDSM database were reported by experienced radiologists providing various BIRADS parameters (density, assessment and subtlety), BIRADS abnormality description and proven pathology. For each abnormality identified, the radiologists draw free form digital curves defining ground truth regions. We use these regions to define squared regions of interest (ROIs) for use as prototypes of mass. Each DDSM case includes additional information such as patient age, date of study and digitization or digitizer's brand.

The DDSM database contains 2,582 mass prototypes including benign and malignant masses. Some of them were located on the border of the mammograms. Consequently, only 2,324 prototypes could be considered, namely, those which might be taken centered in a square without stretching. Some mass prototype examples are shown in Figure 1.

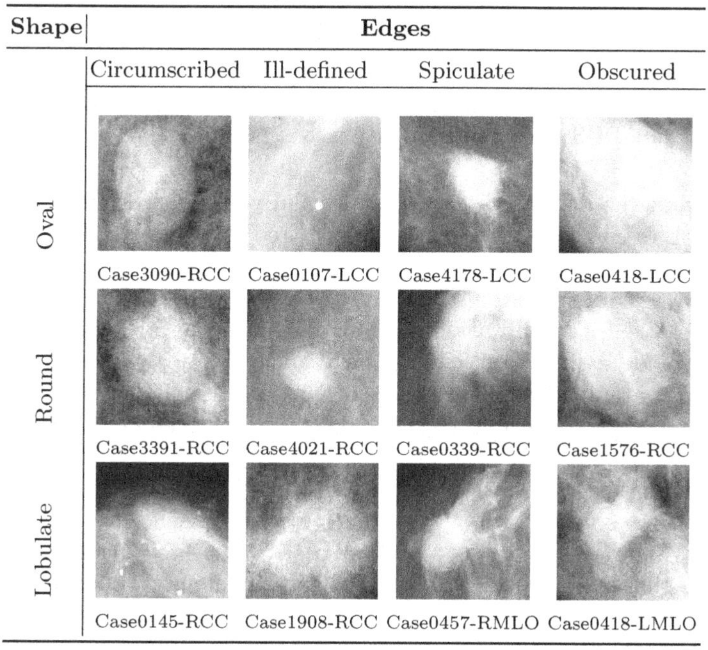

Shape	Edges			
	Circumscribed	Ill-defined	Spiculate	Obscured
Oval	Case3090-RCC	Case0107-LCC	Case4178-LCC	Case0418-LCC
Round	Case3391-RCC	Case4021-RCC	Case0339-RCC	Case1576-RCC
Lobulate	Case0145-RCC	Case1908-RCC	Case0457-RMLO	Case0418-LMLO

Fig. 1. Mass samples for each shape and margin combination. Each ROI has been resized to a common size of 128×128 *pixels*. Mammogram case and view is located under each ROI.

Regions of Interest. Ground truth regions are defined in the database by a chain code which generates a free hand closed curve. We use the chain code to determine the smallest square region of the mammogram that includes the manually defined region. Therefore, if the mass is located near one edge of the mammogram, this procedure may not be able to obtain a squared region from

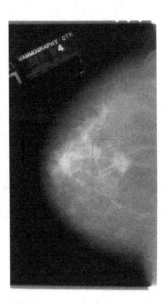

Fig. 2. Ground truth region defined by radiologist (red solid line) and considered ROI (purple box) on a DDSM mammogram

the image and the mass is discarded as a valid prototype. Figure 2 shows the ground truth region coded by the radiologist (red solid line) and the area to be used as ROI (purple box).

USF's DDSM mammograms were digitized with four different scanners for which optical density calibration and spatial resolution are known [2]. Three scanners provide a linear optical response and the fourth one of logarithmic type. To eliminate the dependence of the origin of each digitized mammogram, all obtained ROIs were converted to optical density using the referenced calibration parameters.

The generated regions have different sizes but the selected image feature extractor needs to operate on regions with the same size. So, we need to reduce the size of the selected regions to common sizes. The reduction of ROIs to a common size has demonstrated to preserve mass malignancy information [6,7,8]. To determine the optimum region size, we resized each ROI to two sizes: 32×32, 64×64 pixels. We also tried other sizes such as 128×128 pixels, but the performance obtained with this size was not better than that obtained with the two smaller sizes, whereas the computation time was much greater. Resizing has been carried out using the bilinear interpolation algorithm provided by the OpenCV library [9].

2.2 Independent Component Analysis

The original motivation of ICA is to solve problems known as *blind source separation*(BBS). These problems consist in the following: suppose that we have n

signals. The objective is to develop the signals registered by the sensors $(\mathbf{x}_i)$ as a linear combination of n sources $(\mathbf{s}_j)$, in principle unknown [10].

$$x_i = \sum_{j=1}^{n} a_{ij} s_j$$

The goal of ICA is to estimate the mixture matrix $\mathbf{A} = (a_{ij})$, along with the sources $\mathbf{s}_j$. The ICA model supposes that the observed signals are a linear transformation of hidden sources: $\mathbf{x} = \mathbf{A} \cdot \mathbf{s}$. In general, the mixture matrix $\mathbf{A}$ is invertible, so we have:

$$\mathbf{x} = \mathbf{A} \cdot \mathbf{s} \Rightarrow \mathbf{s} = \mathbf{W} \cdot \mathbf{x} \text{ with } \mathbf{W} = \mathbf{A}^{-1}$$

It is important to remark that:

- The key of ICA estimation is to suppose that hidden sources $(\mathbf{s})$ are non-gaussian and statistically independent.
- We cannot determine the variances (energies) of the independent components.
  - Therefore, the magnitudes of the s_i can be freely normalized.
  - We cannot determine the order of the independent components.

We can use this technique for feature extraction since the components of $\mathbf{x}$ can be regarded as the characteristics representing the objects (patterns) [10]. We have used the FastICA algorithm [11] proceeding as follows:

- We start with N samples (patches, N vectors of dimension p) forming the patches matrix $(\mathbf{x})$ where each row is a patch $(\mathbf{i})$, therefore, the dimensions of this matrix are $N \times p$.
- First, the data are centered by subtracting their averages. That is, to each element is subtracted the mean of its column $(\mathbf{m} = E\{\mathbf{x}\})$ so as to make $(\mathbf{x})$ a zero-mean variable, which implies that $\mathbf{s}$ is zero-mean as well. After estimating the mixing matrix $(\mathbf{A} = (a_{ij}))$ with centered data, we can complete the estimation by adding the mean vector of $\mathbf{s}$ back to the centered estimates of $\mathbf{s}$. The mean vector of $\mathbf{s}$ is given by $\mathbf{A}^{-1}\mathbf{m}$.
- Then a whitening process is applied. This transformation consist in to uncorrelate the data so that their variances are equal to 1 and the covariance matrix is the identity matrix. The dimension of this new whitening matrix is $p \times p$.
- To reduce the size of the input space is applied Principal Component Analysis (PCA) [12], ordering the array of eigenvectors (whitening matrix) by its eigenvalues from highest to lowest and discarding those with lower eigenvalue that will be those with a smaller contribution the variance. Taking q $(q < p)$ first components we obtain the matrix $\mathbf{K}_{PCA}$ of dimension $(q \times p)$.
- Now, taking as input this matrix and applying the ICA algorithm is obtained the ICA transformation matrix of dimension $(q \times q)$.

– Finally considering a new matrix $(\mathbf{W}_T)$, multiplication of the previous two $(\mathbf{W_T} = \mathbf{K^T_{PCA}} \cdot \mathbf{W} \ (p \times q))$, in which each row is a vector of the new base, q characteristics can be extracted of each original input $(\mathbf{i})$ simply by multiplying the matrix $\mathbf{W}_T$ for each of them.

$$\mathbf{c} = \mathbf{i} \cdot \mathbf{W_T}$$

Following this process we can express, as many other transformations (wavelets, Gabor filters, ... [13]), the image (or a image patch) as a linear superposition of some basis functions (basis images in our case) $a_i(x, y)$:

$$I(x, y) = \sum_{i=1}^{p} a_i(x, y) c_i \tag{1}$$

Where the c_i are image-dependent coefficients. This expression is similar to the ICA model, and we can visualize this idea in Figure 3. In that way, estimating a basis images using **ICA**, we could obtain a basis adapted to our data.

Fig. 3. Example ICA basis and expansion of two mass prototypes

In Figure 3, one can see an example of an ICA basis of 20 components that operates on prototypes with dimension 64×64 pixels. In this figure, there are 20 components extracted by the ICA basis for two mass prototypes. Then one can also see its subsequent expansion using the extracted coefficients and the ICA basis of the form shown in Eq. 1. That is, we can decompose or expand our image $(I(x, y))$ by using a base image $(a_i(x, y))$ multiplied by coefficients (c_i) and adding the mean vector $\mathbf{m} = E\{\mathbf{x}\}$ (not showed in the figure).

2.3 Classification Algorithm

The classification algorithm has the work of learning from data. Usually, a model with excessive complexity leads to poor generalization results. In the learning process is convenient to use, at least, two independent sets of patterns: one for training and another for testing. Or, as in this work, we have used three independent sets of patterns: one for training, one for avoid the overtraining and another for testing [14]. We have used Neural Networks and Support Vector Machine (SVM) [15] classifiers.

Neural Networks (NN). We have used the classical feed-forward multilayer perceptron with a single hidden layer and, a variant of Back-Propagation (BP) algorithm named Resilient Back-Propagation (Rprop) [16] to adjust the weights. Rprop is a local adaptive learning scheme performing supervised batch-learning in multilayer perceptron with faster convergence than the standard BP algorithm. The basic principle of Rprop is to eliminate the negative affect of the size of the partial derivative on the update process. As a consequence, only the sing of the derivative is considered to indicate the direction of the weight update [16]. The library of functions of the Stuttgart Neural Network Simulator environment [17] were used to generate and train the NN classifiers. To avoid local minima during the training process each setting was repeated four times changing randomly the initial weights in the net. Furthermore, the number the neurons in the hidden layer could change between 50 and 650 in steps of 50.

Support Vector Machines. The goal of SVM is to find a model (based on the training prototypes) which is able to predict the class membership of the prototypes of the test subset, based on the value of their characteristics.

Given a labeled training set of the form $(\mathbf{x}_i, \mathbf{y}_i)$, $i = 1, \ldots, l$ where $\mathbf{x}_i \in \Re^n$ and $\mathbf{y} \in \{1, -1\}^l$, the SVM algorithm requires the following optimization problem to be solved:

$$min_{w,b,\xi} \; \frac{1}{2} w^T w + C \sum_{i=1}^{l} \xi_i$$
$$where \; y_i \left(w^T \phi\left(\mathbf{x}_i\right) + b \right) \geq 1 - \xi_i,$$
$$\xi_i \geq 0 \tag{2}$$

This algorithm works by projecting the training vectors $\mathbf{x}_i$ onto a higher-dimensional space than the original. The final dimension of this space depends on the complexity of the input space. Thus SVM finds a linear separation by means of a hyperplane with a maximal (i.e., optimal) margin of separation between classes in this higher dimensional space.

The parameter C $(C > 0)$ shown in the model is a penalty term to control the error, and $K(\mathbf{x}_i, \mathbf{x}_j) \equiv \phi(\mathbf{x}_i)^T \phi(\mathbf{x}_j)$ is a kernel function to project the input data onto to a higher dimensional space. We have used LibSVM [18] library in this work with a radial basis function (RBF: $K(x_i, x_j) = \exp(-\gamma \|x_i - x_j\|^2)$, $\gamma > 0$) as kernel function. To find the optimal configuration of the parameters in the algorithm γ could change between -5 and 20 in step of 0.5 and the penalty parameter C between -5 and 10, also, in steps of 0.5.

3   System

In this section, we provide a description of the main steps of our system. The first task is to obtain the prototypes of masses and normal tissue. The prototypes of masses are obtained as was explained in Section 2.1 and the prototypes of normal tissue were selected randomly from the normal mammograms. This normal tissue prototypes were caught originally with sizes that randomly ranging from the smallest to the largest of the sizes found in the DDSM database for masses. Then, applying the FastICA algortithm [11] as is described in Section 2.2 and using *log cosh* function to estimate neg-entropy were obtained the ICA basis (ICA-based feature extractor). To obtain the optimal configuration of the system were generated different ICA basis to extract different number of features (from 10 to 65 in steps of 5) from the original patches and, in addition, operating over patches of different dimensions (as said before, 32×32 and 64×64).

We did a double training process, on the one hand, we trained NN classifiers and, on the other hand, we trained SVM classifiers. After training process were obtained the results shown in Figure 3, where can be seen the results obtained over the test subsets in a 10-fold cross validation scheme. In that way, we find the best configuration of the feature extractor. This study was made with a total of 5,052 prototypes: 1,197 of malignant masses, 1,133 of benign masses and 2,722 of normal tissue. We found that the optimal configuration for the ICA-based feature extractor, for a NN classifier, was a feature extractor that operated on prototypes of 64×64 pixels extracting 10 components (average success 86.33%). And, for a SVM classifier, the best configuration was for a feature extractor that operated also on prototypes of 64×64 pixels extracting 15 components (average success 88.41%). Therefore, the results shown in the following section were obtained using these configurations for the ICA-based feature extractor in each case.

4 Results

In this work, our main interest was to evaluate the robustness of our CAD system. We have included all the prototypes of masses found in the DDSM which could be obtained as a square shape without stretching them by determining the smallest square region that includes the complete ground truth. The distribution of prototypes is shown in Table 1. As can be seen in this table, the number of prototypes on the learning and test sets is quite different depending on the considered scanner, being the most heterogeneous distribution for the DBA M2100 scanner. For this scanner, no prototypes were found of benign masses and the number of prototypes of malignant masses is much lower than the normal tissue prototypes in the learning set. This, as will be seen in the results, is a big handicap to train the classifiers.

In Table 1 one can see that the number of normal tissue prototypes from a DBA scanner is more than half of total normal tissue prototypes. This is due to that in the DDSM database there are 12 volumes of normal mammograms with different number of cases by each volume, with four mammograms by case. From these 12 volumes 6 were digitized with a DBA scanner, among them, those that have a bigger number of cases. Therefore, it seems clear that if we have selected the normal tissue prototypes randomly from the normal mammograms the number of prototypes from a DBA scanner should be, at least, the half of

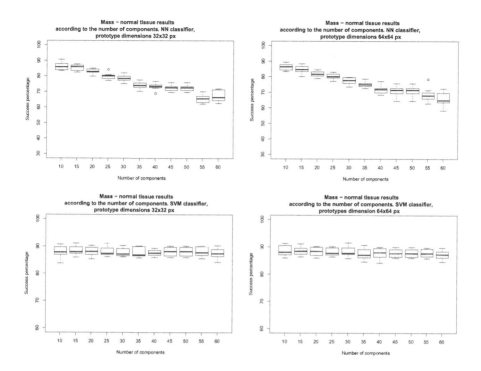

Fig. 4. Choosing the best configuration to the feature extractor. The top row shows the results when we use a NN classifier, while the bottom row shows the results for a SVM classifier. In both cases, prototypes of 32×32 first column and 64×64, second column.

the total of normal tissue prototypes. On the other hand, the number of volumes of *"cancer"* from a DBA scanner is only two and of *"benign"* zero. The number of prototypes is more equilibrated for the rest of scanners.

The results, presented in Table 2, correspond to feature extractors based on ICA described in the previous Section 3 for each classifier. As expected, the most suitable distribution of prototypes is for overall results, since, as seen in Table 1, the number of prototypes in the learning set is much greater than the number of prototypes in the test set. And, in addition, the trained classifier with this distribution can *"learn"* prototypes from all the scanners, while, the trained classifiers with the other distributions only can learn prototypes from one scanner. Anyway, in theory this should not affect to the results, because the prototype images are transformed to optical density by using the scanner calibration parameters provided by the USF's DDSM database authors.

In Table 2 one can see that when the number of prototypes in the learning set is low (HOWTEK scanners) the performance with SVM classifiers is a little better than with NN classifiers. This seems to agree with [19] where was made a comparison of the performance and robustness of different types of classifiers in different settings. In contrast, when the number of prototypes is large enough

Table 1. Distribution of prototypes in learning and test sets, depending on the used scanner and without considering the scanner (overall results). Along with the scanner's name appears the identification of the institution in which they were used.

Distribution of prototypes depending on the scanner.			
Scanner	**Pathology**	**Learning**	**Test**
HOWTEK 960 (MGH)	Malignant	345	851
$43.5\mu m/pixel$	Benign	485	648
linear calibration	Normal	312	2410
HOWTEK MultiRAD 850 (WU)	Malignant	107	1089
$43.5\mu m/pixel$	Benign	154	979
linear calibration	Normal	417	2305
DBA M2100 (MGH)	Malignant	105	1091
$42\mu m/pixel$	Benign	0	1133
logarithmic calibration	Normal	1668	1054
LUMISYS 200 laser (WFU & SH)	Malignant	639	557
$50\mu m/pixel$	Benign	494	639
linear calibration	Normal	325	2397
	Malignant	1074	122
Overall	Benign	1032	101
	Normal	2440	282

Table 2. In this table the obtained results are shown. The learning set is divided into two: the train subset and the validation (Val) subset, corresponding to the first the 80% and the second the 20% of the prototypes in the learning set.

Success results for discriminating Mass-Normal tissue					
	SVM classifier			**NN classifier**	
Scanner	**Learning (%)**		**Test(%)**	**Learning(%)**	**Test(%)**
	Train(80%)	Val(20%)		Train(80%) Val(20%)	
HOWTEK 960	96.38	93.89	75.29	95.40 89.52	71.22
HOWTEK MD850	98.71	91.91	82.71	87.27 80.88	80.56
DBA M2100	99.80	97.46	58.69	99.29 96.90	57.75
LUMISYS	94.17	90.75	71.72	94.68 88.01	76.62
Overall results	92.65	88.20	88.41	90.73 88.37	86.33

(LUMISYS scanner), the performance with NN classifiers seems to be a little better than with SVM classifiers. However, considering overall results, the performance is slightly better with SVM classifiers, which contradicts the previous statement. Finally, when the number of prototypes is not enough (DBA scanner) the performance for both classifiers is quite bad.

5 Conclusions

This study about the robustness of our CAD system has been made using a very heterogeneous dataset. Data (mammograms and reports) comes from four clinical sites. The mammograms were digitized in three different institutions using, in principle, four different scanners but the DBA scanner, used at MGH, was retired due to continuing performance difficulties [20]. We can evaluate the performance obtained in each case choosing as a reference the overall performance over the test subset, because in this setting the learning set was formed by prototypes of all scanner and the classifiers can *learn* prototypes of all them. Considering a SVM classifier the differences in performance were: 14.8% for Howtek 960, 6.4% for Howtek MD850, 33.6% for DBA and 18.8% for Lumisys. And considering a NN classifier, 17.5% for Howtek 960, 6.6% for Howtek MD850, 33.1% for DBA and 11.6% for Lumisys. Taking into account all was said about the DBA scanner, we have to say that the results for this scanner are not conclusive. On the other hand, the least variation in the performance is found for Howtek MD850 scanner for both classifiers which could indicates that for this scanner had a good representation of the entire dataset. For Howtek 960 the performance varitation is higher for a SVM classifier and for Lumisys the performance variation is higher for a NN classifier. Here, we can see that when the number of prototypes in the learning set is large enough, the NN classifiers obtain a better generalization capabiltiy than the SVM classifiers. While, when the number of prototypes in the learning set is low the generalization capability of the SVM classifiers is better than with NN classifiers.

Acknowledgments. This work has been partly supported by "Junta de Extremadura" and FEDER through projects PRI08A092, PDT09A036, GR10018 and PDT09A005.

References

1. Serio, G.V., Novello, A.C.: The advisability of the adoption of a law that would expand the definition of mammography screening to include the review of x-ray examinations by use of a computer aided detection device. Technical report, The Superintendent of Insurance in consulation with the Commissioner of Health, USA (2003)
2. Heath, M., Bowyer, K., Kopans, D., Moore, R., Kegelmeyer, P.: The digital database for screening mammography. In: Proceedings of the 5th International Workshop on Digital Mammography, pp. 212–218 (2000)
3. Horsch, A., Hapfelmeier, A., Elter, M.: Needs assessment for next generation computer-aided mammography reference image databases and evaluation studies. International Journal of Computer Assisted Radiology and Surgery, 1–19 (2011)
4. Hyvärinen, A., Hurri, J., Hoyer, P.O.: Natural Image Statistics. A Probablisctic Approach to Early Computational Vision. Springer (2009)
5. Anjali, P., Ajay, S.: A review on natural image denoising using independent component analysis (ICA) technique. Advances in Computational Research 2(1), 06–14 (2010)

6. Campanini, R., Dongiovanni, D., Iampieri, E., Lanconelli, N., Masotti, M., Palermo, G., Riccardi, A., Roffilli, M.: A novel featureless approach to mass detection in digital mammograms based on support vector machines. Physics in Medicine and Biology 49, 961–975 (2004)
7. Angelini, E., Campanini, R., Iampieri, E., Lanconelli, N., Masotti, M., Roffilli, M.: Testing the performances of different image representations for mass classification in digital mammograms. International Journal of Modern Physics C [Computational Physics and Physical Computation] 17(1), 113–131 (2006)
8. Hong, B.-W., Brady, J.M.: A Topographic Representation for Mammogram Segmentation. In: Ellis, R.E., Peters, T.M. (eds.) MICCAI 2003. LNCS, vol. 2879, pp. 730–737. Springer, Heidelberg (2003)
9. Bradski, G., Kaehler, A.: Learning OpenCV: Computer Vision with the OpenCV Library. O'Reilly Media (2008)
10. Hyvärinen, A., Karhunen, J., Oja, E.: Independent Component Analysis. Adaptive and Learning Systems for Signal Processing, Communications, and Control. John Wiley & Sons (2001)
11. Ripley, B.: FastICA Algorithms to perform ICA and Projection Pursuit (February 2009), http://cran.r-project.org/web/packages/fastICA/fastICA.pdf
12. Jolliffe, I.T.: Principal Component Analysis, 2nd edn. Series in Statistics. Springer (2002)
13. Gonzalez, R.C., Woods, R.E.: Digital image processing, 3rd edn. Prentice-Hall, Upper Saddle River (2008)
14. Bishop, C.M.: Pattern Recognition and Machine Learning. Springer (2006)
15. Vapnik, V.N.: The Nature of Statistical Learning Theory, 2nd edn. Statistics for Engineering and Information Science. Springer (2000)
16. Riedmiller, H., Braun, H.: A direct adaptive method for faster backpropagation learning. The RPROP algorithm. In: IEEE International Conference on Neural Networks, pp. 586–591 (1993)
17. Zell, A., Mache, N., Huebner, R., Mamier, G., Vogt, M., Schmalzl, M., Herrmann, K.: SNNS (Stuttgart Neural Network Simulator). In: Neural Network Simulation Environments, pp. 165–186 (1994)
18. Chang, C.C., Lin, C.J.: LIBSVM: A library for support vector machines. ACM Transactions on Intelligent Systems and Technology 2(3), 27:1–27:27 (2011)
19. Lausser, L., Kestler, H.A.: Robustness Analysis of Eleven Linear Classifiers in Extremely High–Dimensional Feature Spaces. In: Schwenker, F., El Gayar, N. (eds.) ANNPR 2010. LNCS, vol. 5998, pp. 72–83. Springer, Heidelberg (2010)
20. Bowyer, K.W.: Digital Image Database with Gold Standard and Performance Metrics for Mammographic Image Analysis Research. Technical report, U.S. Army Medical Research and Materiel Command Fort Detrick, Maryland 21702-5012 (August 1999)

Facial Expression Recognition Using Game Theory

Kaushik Roy and Mohamed S. Kamel

Centre for Pattern Analysis and Machine Intelligence
University of Waterloo, ON, Canada
kaushik.roy@uwaterloo.ca

Abstract. Accurate detection of lip contour is important in many application areas, including biometric authentication, human computer interaction, and facial expression recognition. In this paper, we propose a new lip boundary localization scheme based on Game Theory (GT) to improve the facial expression detection performance. In addition, we use GT for selecting the proper set of facial features. We apply the Extended Contribution-Selection Algorithm (ECSA) for the dimensionality reduction of the facial features using a coalitional GT-based framework. We have conducted several sets of experiments to evaluate the proposed approach. The results show that the proposed approach has achieved recognition rates of 93.1% and 92.7% on the JAFFE and CK+ datasets, respectively.

Keywords: Facial expression recognition, coalitional game theory, extended contribution selection algorithm.

1   Introduction

Automatic lip boundary detection plays an important role in numerous application areas, including biometric authentication, human computer interaction, and facial expression recognition (FER) [1]. However, extracting the lip boundary from the mouth region is quite complicated and difficult due to large variations emerged from different speakers, illumination conditions, poor texture of lips, weak contrast between lip and skin, high deformability of lip, wrinkles, and occlusions by beard and moustache. Various methodologies for lip contour detection have been published during the last few years [2]. The state-of-the-art lip contour detection methods can be divided into three broad categories: 1) model-based, 2) color/gray level analysis-based, and 3) level set-based approaches. The model based approaches may not provide satisfactory performance due to poor contrast between lip and skin color. The color/gray level analysis-based approaches are computationally efficient. However, these methods result in large color noise and are sensitive to color contrast. The basic idea behind the level set-based approaches is to represent the lip contours as the zero level set of an implicit function defined in a higher dimension, usually referred as the level set function, and to evolve the level set function according to a partial differential equation (PDE). The level set-based methods mainly depends on the image gradients and thus, are highly sensitive to the presence of noise and poor image

N. Mana, F. Schwenker, and E. Trentin (Eds.): ANNPR 2012, LNAI 7477, pp. 139–150, 2012.

contrast. In this paper, we propose a lip localization approach based on Game Theory (GT) to track the lip boundary from the mouth region. We propose to apply a parallel game-theoretic decision making procedure by modifying Chakraborty and Duncan's algorithm [3], which combines (1) the region-based and gradient-based boundary finding methods, and (2) integrates the complementary strengths of each of these individual methods. This GT-based scheme is robust to noise and poor localization and performs well against weak lip/skin boundaries. Previous work on FER has focused mainly on the issues of feature extraction and facial pattern classification. However, less effort has been given to the critical issue of feature selection. In this paper, we propose a two-stage feature selection scheme, which focuses on feature ranking and redundancy reduction. First, we apply Information Gain (IG) to rank the best textural features [4]. Though IG has been regarded as one of the most efficient measures of feature ranking in classification problem, it has a serious weakness of ignoring the redundancy among higher ranked features. Therefore, finally, we propose a feature reduction scheme in the context of GT [5]. The GT-based feature selection scheme is generally quite effective in a rapid global search of large, non-linear and poorly understood spaces. It also suggests a particularly attractive approach in solving the problem occurred due to highly dimensionality of facial features with small sample size. The game-theoretic scheme takes into account the correlation of features while reducing the dimensionality. An iterative algorithm for feature ranking, called the contribution-selection algorithm [5], is enhanced, called Enhanced Contribution-Selection Algorithm (ECSA) hereafter, and used to select the optimal feature subset. This algorithm depends on the Multi-perturbation Shapley analysis (MSA), a framework that is based on the coalitional GT, to estimate the usefulness of features and rank them accordingly.

2   Lip Localization

In this section, we mainly focus on GT-based lip boundary detection approach. First, we apply Viola-Jones method [6] to detect the presence of a human face. An Active Shape Model with Local Features (ASMLF) [7] is applied to find a set of 60 facial feature points from the detected face image. The selected set of feature points is used to delineate the regions of interest. We apply the H-minima transform on the complemented input image to reduce all minima in the detected mouth image whose depth is less than a threshold. To enhance the quality of the image and to suppress the effect of noise, a 2D Wiener filter is deployed. Watershed transformation is used to divide the filtered image into several catchment basins, which consist of its own regional minimum. The pixels of final outcome are labelled according to a specific catchment basin number. [2]. This process is shown in Figure 1. Finally, we apply a parallel game-theoretic decision making procedure by modifying the Chakraborty and Duncan's algorithm [3], which combines the region-based segmentation and the boundary finding methods for the optimal estimation of lip boundary. The game is usually played out by a set of decision makers (or "players"), which in our case, corresponds to the two segmentation schemes, namely, the region-based and the

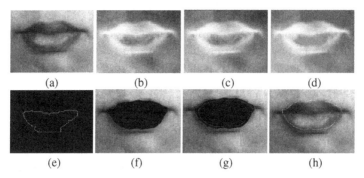

(a)   (b)   (c)   (d)

(e)   (f)   (g)   (h)

Fig. 1. Lip contour detection using GT. (a) Delineated mouth region from an original face image. (b) Complemented image. (c) Image after H-minima transform. (d) Enhancement using 2D Weiner filter. (e) Image after watershed segmentation. (f) Segmented area filled-up with black pixels. (g) Contour initialization. (h) Final contour.

gradient-based segmentation methods [3]. The lip segmentation problem can be formulated as a two-player game. If p^1 represents the set of strategies of the Player 1, and p^2 denotes the set of strategies of the Player 2, then each player tries to minimize the payoff function, $F^i(p^1, p^2)$. The main objective is to find the *Nash Equilibrium* (NE) of the system $(\bar{p}^1, \bar{p}^2)$, such that:

$$F^1(\bar{p}^1, \bar{p}^2) \leq F^1(p^1, \bar{p}^2), F^2(\bar{p}^1, \bar{p}^2) \leq F^2(\bar{p}^1, p^2) \qquad (1)$$

We move toward the NE iteratively by taking t as the time index and can formulate the game as follows:

$$p^1_{t+1} = \underset{p^1 \in P^1}{\arg\min} F^1(p^1, p^2_t); \; p^2_{t+1} = \underset{p^2 \in P^2}{\arg\min} F^2(p^1_t, p^2) \qquad (2)$$

In [3], Chakraborty and Duncan proved that there is always an existing NE solution if F^1 and F^2 are of the following form [3]:

$$F^1(p^1, p^2) = f_1(p^1) + \alpha f_{21}(p^1, p^2) \qquad (3)$$

$$F^2(p^1, p^2) = f_2(p^2) + \beta f_{12}(p^1, p^2) \qquad (4)$$

where α and β are scaling constants, F^i is bounded in $p^i \in P^i$, F^i is continuously second-order differentiable in $p^i \in P^i$, and there is an existing closed neighborhood of $u^i \subseteq p^i$ such that F^i is strongly convex in u^i. In the region-based method, the image is partitioned into connected regions by grouping the neighbouring pixels of similar intensity levels. The adjacent regions are then merged under some criteria involving the homogeneity or sharpness of the region boundaries. Now, if $y_{i,j}$ is the intensity of a pixel at (i, j) of the original image and $x_{i,j}$ is the intensity of a pixel at (i, j) of the segmented image, then, a common approach is to minimize an objective function of the form [3]:

$$E = \sum_{i,j}(y_{i,j} - x_{i,j})^2 + \lambda^2 (\sum_{i,j} \sum_{i_s, j_s}(x_{i,j} - x_{i_s, j_s})^2) \qquad (5)$$

where, i_s and j_s are indices in the neighborhood of pixel $x_{i,j}$, and λ is a constant. In the above equation, the first term on the right-hand side represents the data fidelity term, and the second term on the right-hand side enforces the smoothness. To find the lip contour, the objective functions are described as follows:

For the region-based module (Player 1):

$$F^1(p^1, p^2) = \underset{x}{min} \left[\Sigma_{i,j}(y_{i,j} - x_{i,j})^2 + \lambda^2 \left(\Sigma_{i,j}(x_{i,j} - x_{i-1,j})^2 + \Sigma_{i,j}(x_{i,j} - x_{i,j-1})^2 \right) \right] + \alpha \left[\Sigma_{i,j \in A_{\vec{p}}}(x_{i,j} - u)^2 + \Sigma_{i,j \in \bar{A}_{\vec{p}}}(x_{i,j} - v)^2 \right] \qquad (6)$$

where, $y_{i,j}$ is the intensity of the original image, $x_{i,j}$ is the intensity of the segmented image given by p^1 as mentioned earlier, u is the intensity inside the contour given by p^2, and v is the intensity outside the contour given by p^2. $A_{\vec{p}}$ corresponds to the points that lie inside the contour, and $\bar{A}_{\vec{p}}$ represents those points that lie outside the contour. The first term on the right-hand side of (6) minimizes the difference between the pixel intensity values and the obtained region, as well as enforces continuity. The second term tries to match the region and the contour. In the region growing approach, we select an initial area within the region of interest for the lip boundary detection. At each iteration, the neighbouring pixels are observed and the value of E is measured from (5). The pixels, for which the value of E is less than a predefined threshold, are accepted into the region. The objective function of the Player 2 (i.e., the boundary finding module) is as follows:

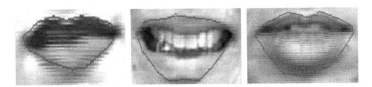

Fig. 2. Lip contour detection using GT on CK+ database

$$F^2(p^1, p^2) = \underset{\vec{p}}{argmax} \left[M_{gradient}(I_g, \vec{p}) + \beta M_{region}(I_r, \vec{p}) \right] \qquad (7)$$

where, $\vec{p}$ denotes the parameterization of the contour given by p^2, I_g is the gradient image, I_r is the region segmented image, and β is a constant. In this paper, we apply a Variational Level Set (VLS)-based active contour model to parameterize and represent the lip contour data during the game-theoretic propagation [8]. Figure 2 shows the lip segmentation results. For feature extraction, we apply a discrete set of 1D log-Gabor kernels, which contains 4 spatial frequencies and 6 different orientations from 0° to 180°, differing in 30 steps that makes a filter bank of 24 different Gabor filters. These log-Gabor filters are deployed to each of the images and filter responses are obtained only at the selected fiducial points [9]. Therefore, the facial expressions in an input image are represented by a feature vector of length 1440 elements (60 fiducial points, 24 filter responses per point).

3   Feature Ranking and Reduction

We propose a two-stage feature selection strategy, including featuring ranking and removing the redundant terms. The IG has been regarded as one of the most efficient measures of feature ranking in classification problem [4]. Therefore, in the first stage, we apply a feature ranking measure based on IG to select the most informative and relevant features. The IG can be calculated as [4]:

$$
\begin{aligned}
InG(x_j) = \\
- \sum_{k=1}^{C} P(y_k). log P(y_k) + \\
P(x_j) \sum_{k=1}^{C} P(y_k|x_j). \log P(y_k|x_j) + P(\bar{x}_j) \sum_{k=1}^{C} P(y_k|\bar{x}_j). \log P(y_k|\bar{x}_j)
\end{aligned}
\tag{8}
$$

where $P(y_k)$ is the probability of a sample feature vector belonging to the class y_k, $P(x_j)$ is the probability of a sample feature vector containing the feature x_j, $P(y_k|x_j)$ is the conditional probability of y_k given the feature x_j. The number of class is denoted by C. In our application, the extracted feature set of 1440 elements from each facial image is ranked by IG. In the second stage, we apply a redundancy reduction algorithm in the context of cooperative or coalitional games, a notion from GT [5]. The algorithm is based on the MSA [5], a framework which relies on GT to estimate the effectiveness of features. This algorithm iteratively computes the usefulness of features and selects them accordingly using forward selection process.

An iterative algorithm for the feature selection, called ECSA, is used to optimize the performance of the classifier on unseen data [5]. The ECSA algorithm combines both the filter and wrapper approaches. However, unlike the filter methods, the features are ranked at each step by using the classifier as a black box. The ranking is based on the *Shapley value* [5], a well known concept from GT, to estimate the importance of each feature for the task at hand by taking into account the interactions between the features. Formally, we can define the coalitional game theory by the pair *(N, u)*, where $N = \{1, 2, \ldots \ldots, n\}$ is the set of all players and $u(F)$, for every $F \subseteq N$, denotes a real number associating a value with the coalition F. GT represents the contribution of each player to the game by constructing a certain value function, which assigns a real-value to each player and the values correspond to the contribution of the players in achieving an optimal payoff. The calculation of the contribution value is based on the Shapley value [5]. Essentially, the Shapley value of a player is a weighted mean of its marginal value, averaged over all the possible subsets of players. If we transform the concept of GT into the arena of facial feature subset selection, in which the contribution of each feature is estimated to generate a classifier, the players N are mapped to the features of a dataset and the payoff is denoted by a real valued function *u(F)*. We can calculate the contribution values from the sampled permutations of the whole set of players, with d being the bound on the permutation-size:

$$
\theta_i(u) = \frac{1}{|\Pi_d|} \sum_{\pi \in \Pi_d} \Delta_i(F_i(\pi))
\tag{9}
$$

where Π_d denotes the set of sampled permutations on subsets of size d. The ECSA is iterative in nature, and can either adopt a forward selection or backward elimination approach. In this paper, we consider only the Forward Feature Selection (FFS) approach for the GT-based framework. Based on the contribution value, ECSA ranks each feature and then selects features with the lowest contribution. The algorithm continues to calculate the contribution values of the remaining features, given those that have already been selected, and selects the new features, until the contribution values of all the candidate features fall below a contribution threshold. The algorithm can be regarded as a generalization of filter methods. However, the main idea of the algorithm is that the contribution value is calculated for each feature according to its assistance in improving the classifier's performance, which is generated using a specific induction algorithm, and in conjunction with other features. We propose the following payoff function that minimizes the within-class distance and maximizes the between-class distance:

$$min\big(u(F)\big) = min(W_l) + min(1/(B_l + 1)) \qquad (10)$$

In (10), W_l denotes the within-class distance and can be defined as

$$W_l = \frac{1}{n_l} \sum_{k=1}^{n_l} (X_k^l - m_l)^T (X_k^l - m_l) \qquad (11)$$

where $l = 1,2,3, \dots \dots, c$ and m_l is the mean vector of class l. B_l in (10) denotes the between-class distance and can be defined as

$$B_l = \sum_{l=1}^{c} P_l \times (m_l - m)^T (m_l - m) \qquad (12)$$

where m is the mean vector of all samples. The ECSA algorithm in its FFS version is depicted as follows:

Extended Contribution Selection Algorithm (P, T_d, d, f):
P is the set of input features, T_d denotes the contribution threshold, d is the maximal permutation size for estimating contribution values, and f is the selected feature subset size in each phase. This algorithm calculates the contribution value of feature, p using the payoff function, $u(F)$ described above, and selects at most f features with the lowest contribution values that fall below T_d.

1. *SelectedFeatures $:= \phi$*
2. *For each $p \in P \setminus SelectedFeatures$*
   i) *$ECNT_p = Contributions\ (p, SelectedFeatures, d)$*
3. *if min $ECNT_p < T_d$*
   i) *SelectedFeatures $:=$*
      SelectedFeatures $\cup$ Selection $(\{ECNT_p\}; f, T_d)$
   ii) *Goto Step 2*
      else
   iii) *return SelectedFeatures*

The case *SelectedFeatures $:= \phi$* is handled by returning the fraction of majority class instances. In our feature selection algorithm, we use the Shapley value heuristically to estimate the contribution value of a feature. The decision tree is used for the feature selection, and the SVM is deployed to perform the actual prediction on the selected features based on the performance of the classifier [10].

4   Performance Evaluation

Extensive experiments were conducted on the following two databases, namely, JAFFE [11], and Cohn-Kanade Version 2 or Ck+ [12]. Figure 3 exhibits the results of facial feature tracking using the ASMLF with the game-theoretic approach. From the Figure 3, we can find that the applied game-theoretic approach with the local features accurately finds the feature points on both of the databases. The GT-based lip contour detection process, further, enhances the feature tracking performance. An extensive set of experiments was conducted on all the datasets, and the coupling coefficients, α and β were set to 0.27 when the game-theoretic integration module was used. To obtain the contour data of the lip boundary during game-theoretic evolution, the selected parameter values using the VLS algorithm were set to $\mu = 0.001, v = 2.0, \lambda = 5.0$ and time step $\tau = 3.0$. The number of models in the shape models was 21 and 35 for the JAFFE and CK+ databases, respectively. To show the effectiveness of the lip contour detection process, we compared the proposed GT-based lip contour detection scheme with Geodesic Active Contour (GAC) [13], ASM proposed by Cootes et al. [14] and ASMLF [7] as shown in Figure 4. The GT-based scheme substantially improves the accuracy of lip detection, especially when mouths are open. The reason is that our proposed scheme uses the region-based information

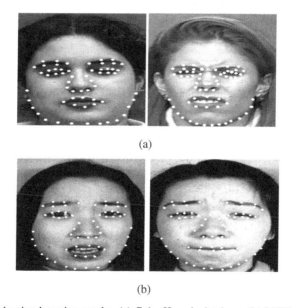

(a)

(b)

Fig. 3. Facial point detection results. (a) Cohn Kanade database. (b) JAFFE database.

as well as the gradient data with the game-theoretic fusion method for lip localization. In the first stage of feature selection, we applied IG to rank the feature from the extracted set of 1440 log-Gabor elements. Fig. 5 shows the accuracy of the selected feature subsets with a different number of top-ranked features obtained using IG on two datasets. Fig.5 (a) shows that IG achieves a reasonable accuracy of 93.80% when the number of selected features is 640. Therefore, a selected set of 640 top-ranked feature elements is used as input to the GT-based feature selection criteria for the further improvement of FER. Similarly, we can find from Fig. 5(b) that an accuracy accuracy of 93.25% is reached when the length of the feature vector is 800. After obtaining 800 top-ranked features from IG, we input them to the GT for further reduction of feature size. The proposed cooperative GT-based feature selection approach is used to reduce the feature dimension without compromising the recognition accuracy. The ECSA is prone to overfitting on the validation set. The

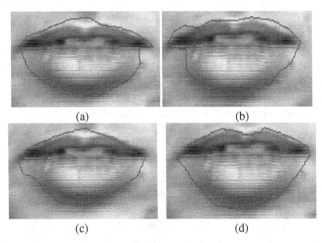

(a)   (b)

(c)   (d)

Fig. 4. Comparison of different lip contour detection methods. (a) GAC, (b) ASM, (c) ACMLF, and (d) ASMLF with GT.

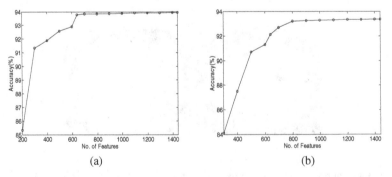

(a)   (b)

Fig. 5. Feature ranking using information gain. (a) JAFFE database. (b) CK+ database.

"curse of dimensionality" appears, and the irrelevant features are selected if the classifier's performance is evaluated on a small validation set. Since the number of samples from most facial expression research is limited, the cross-validation procedure is commonly used to evaluate the performance of a classifier. For the JAFFE and CK+ datasets, we use a 4-fold cross validation to obtain the validation accuracy. Figs. 6 shows the cross-validation accuracies of the selected feature subsets for the FFS scheme on two datasets. From Fig. 6, we can see that the reasonable accuracy is obtained with the FFS scheme when the number of selected features is (a) 313 in JAFFE dataset and (b) 410 in CK+ datasets. In order to show the effectiveness of the proposed feature selection scheme, we compared our GT-based method with Genetic Algorithm (GA) [15] and Mutual Information (MI) [4]. For GA, we deployed the same fitness function used in the current study. Fig. 7 clearly demonstrates that our GT-based scheme outperforms the other two methods in terms of dimensionality reduction, especially when the feature vector length is comparatively small. We can see from Fig. 7(a) that GT achieves an accuracy of 92.81% when the length of feature vector is 410, while GA reaches a similar accuracy level when the feature vector

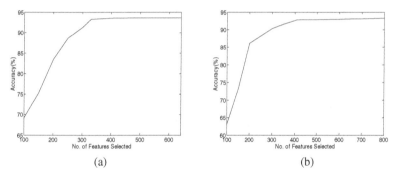

(a)   (b)

Fig. 6. Feature selection using GT with FFS. (a) JAFFE database. (b) CK+ database.

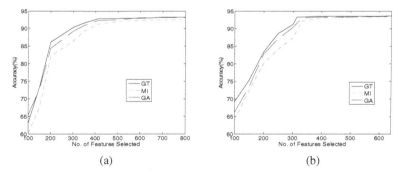

(a)   (b)

Fig. 7. Comparison of different feature selection methods. (a) CK+ database. (b) JAFFE database.

length is 560. However, MI was outperformed by GT and GA in terms of both the accuracy and dimensionality. Similarly, in Fig. 7(b), GT shows an accuracy of 93.20% when the selected size of feature vector is 313, whereas the similar recognition accuracy can be reached by GA and MI when the feature vector sizes are 560 and 595, respectively. We used SVM to classify 7 categories of expressions, namely, neutral, happiness, sadness, surprise, anger, disgust, and fear. In order to analyze the performance of the classifier in recognizing individual expression, a training set of 70 images is produced from the JAFFE database. The training set

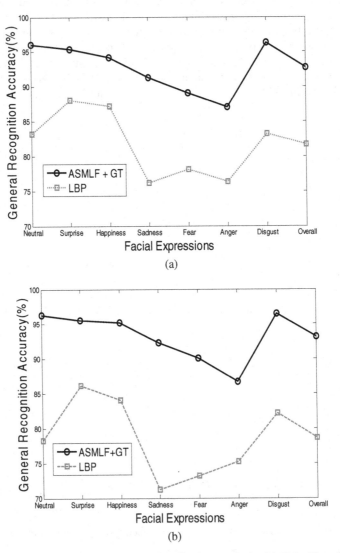

Fig. 8. Comparison of different facial shape localization methods. (a) Cohn-Kanade database. (b) JAFFE database.

contains 7 images for each of the 10 expresser, one image per expression. The other images are then used for testing. For CK+ database, we used a set of 120 facial images for training and the rest of the images were used for testing the overall accuracy of the proposed FER system. We also compared the performance of the proposed algorithm with other existing FER algorithms. Fig. 8 shows the comparison of the proposed algorithm with LBP [1]. We can find that the proposed shape guided approach with the GT-based lip contour detection process shows a better feature point tracking performance than the LBP scheme, since our proposed lip contour detection process enhances the feature point detection accuracy around the mouth region. We can find from Fig. 8 that our method outperforms the conventional LBP method with the accuracies of 93.1% and 92.7% on the JAFFE and CK+ datasets, respectively. Table 1 also demonstrates the comparison of our method with LBP and Boosted LBP reported in [1], and also with PCA-based approach proposed in [8] on JAFFE dataset. From Table 1, we can find that our classification rate on the JAFFE dataset outperforms the other techniques with an accuracy of 93.12%.

Table 1. Comparisons of different FER methods on JAFFE database

Methods	7-Class recognition (%)
LBP [1]	78.6
Boosted-LBP+SVM [1]	81.0
PCA [9]	87.51
Proposed	93.12

5   Conclusions

In this research effort, we have achieved two performance goals. First, a game-theoretic lip contour detection scheme is deployed to enhance the performance of lip localization method. The GT-based algorithm brings together the region-based and boundary-based methods and operates different probability spaces into a common information-sharing framework. The proposed algorithm localizes the lip boundaries from the delineated mouth regions that may be affected by low image intensities, poor acquisition process, opening of the mouth, variability in speaking style, teeth, wrinkles, and by the occlusions of beard and moustache. The localization scheme based on GT avoids the over-segmentation and performs well against the blurred outer lip/skin boundary. Experiments results show that the improved lip localization scheme can enhance the accuracy of the overall facial shape tracking process. Second, a two-stage feature selection approach based on IG and GT is used to find the subset of informative texture features. Further analysis of our results indicates that the proposed feature selection framework is capable of removing redundant and irrelevant features, outperforming other traditional approaches like GA and MI. We validate the proposed FER scheme on JAFFE and CK+ datasets with an encouraging performance.

References

1. Shan, C., Gong, S., McOwan, P.: Facial expression recognition based on Local Binary Patterns: A comprehensive study. Image and Vis. Comput. 27(6), 803–816 (2009)
2. Li, K., Wang, M., Liu, M., Zhao, A.: Improved level set method for lip contour detection. In: IEEE Intl. Conf. Image Process., pp. 673–676 (2010)
3. Chakraborty, A., Duncan, J.: Game-theoretic integration for image segmentation. IEEE Trans. Pattern Anal. and Machine Intell. 21(1), 12–30 (1999)
4. Makrehchi, M., Kamel, M.: Aggressive feature selection by feature ranking. In: Liu, H., Motoda, H. (eds.) Computational Methods of Feature Selection, pp. 313–330. Chapman and Hall/CRC Press (2007)
5. Cohen, S., Dror, D., Ruppin, E.: Feature selection via coalitional game theory. Neural Computa. 19, 1939–1961 (2007)
6. Viola, P., Jones, M.: Robust Real-Time Face Detection. Intl. J. Comp. Vis. 57(2), 137–154 (2004)
7. Ginneken, B., Frangi, A., Staal, J., Romeny, B., Viergever, M.: Active shape model segmentation with optimal features. IEEE Trans. Medical Imaging 21(8), 924–933 (2002)
8. Li, C., Xu, C., Gui, C., Fox, M.: Level set evolution without re-initialization: a new variational formulation. In: Proc. IEEE Intl. Conf. Comp. Vis. and Pattern Recog., pp. 430–436 (2005)
9. Bashyal, S., Venayagamoorthy, G.: Recognition of facial expressions using Gabor wavelets and learning vector quantization. Intl. J. Engg. App. of Artificial Intell. 21(7), 1–9 (2008)
10. Vapnik, V.: Statistical Learning Theory. John Wiley & Sons, New York (1998)
11. Lyons, M., Budynek, J., Akamatsu, S.: Automatic classification of single facial images. IEEE Trans. Pattern Anal. and Machine Intell. 21(12), 1357–1362 (1999)
12. Lucey, P., Cohn, J., Kanade, T., Saragih, J., Ambadar, Z.: The extended Cohn-Kanade dataset (CK+): A complete dataset for action unit and emotion-specified expression. In: IEEE Intl. Conf. Computer Vis. and Pattern Recog. Workshop, pp. 94–101 (2010)
13. Paragios, N., Deriche, R.: Geodesic active contours and level sets for the detection and tracking of moving objects. IEEE Trans. Pattern Analysis and Machine Intelligence 22(3), 266–280 (2000)
14. Cootes, T., Taylor, C., Cooper, D., Graham, J.: Active shape models—their training and application. Computer Vis. Image Understand. 61(1), 38–59 (1995)
15. Goldberg, D.: Genetic algorithms in search, optimization, and machine learning. Addison-Wesley Professional (1989)

Classification of Segmented Objects through a Multi-net Approach

Alessandro Zamberletti, Ignazio Gallo, Simone Albertini, Marco Vanetti, and Angelo Nodari

University of Insubria
Dipartimento di Scienze Teoriche ed Applicate
via Mazzini 5, Varese, Italy

Abstract. The proposed model aims to extend the MNOD algorithm adding a new type of node specialized in object classification. For each potential object identified by the MNOD, a set of segments are generated using a *min-cut* based algorithm with different seeds configurations. These segments are classified by a suitable neural model and then the one with higher value is chosen, in agreement with a proper energy function. The proposed method allows to segment and classify each object simultaneously. The results showed in the experiment section highlight the potential and the cost of having unified segmentation and classification in a single model.

Keywords: object segmentation, object classification, neural networks, minimum cut.

1 Introduction

Many computer vision problems can be reconducted to the identification and recognition of an object as belonging to a class of objects. In some cases, however, it is fundamental to segment the object before classifying it, in order to be able to extract the relevant characteristics from the object. Given an image that can contain objects belonging to different classes, the generic problem of class segmentation consists in the prediction of the membership class of each pixel, or to map a pixel to the "background" if the pixel does not belong to one of the given classes. The most common approach applied for the object segmentation is based on sliding windows [1,2]; other recent works try to solve the problem by classifying a set of segments through a bottom-up approach, towards the classification of the entire object [3,4]. In this work we want to join the two techniques aforementioned in a single model, with the aim to use both methods in a synergistic way.

The Multi-Net for Object Detection (MNOD) [2] is an algorithm which relies on multiple neural networks to deal with the object detection problem, through an image segmentation approach. This model is organized as a directed acyclic graph where each internal node is a neural network: the source nodes extract the features from the input image, and the output of a node becomes the input of

N. Mana, F. Schwenker, and E. Trentin (Eds.): ANNPR 2012, LNAI 7477, pp. 151–162, 2012.

another neural network in a feedforward manner. The output produced by the last node is the final result of the whole algorithm.

The concept that underlies this model is that the prediction performed by a node, given to the following node, should lead to an improvement as if it were a refinement process, and also leads to an improvement in the generalization ability of the whole system, similarly to [5].

The goal of this work is the realization of an algorithm able to segment and classify all the objects of interest present in an image. The segmentation maps produced by the MNOD are used in order to detect several regions of interest (ROI's) inside the processed image. They are detected following a pyramidal approach, in order to maximize the probability that each object of interest is completely surrounded by a ROI boundary. The object in each ROI is segmented using the Min-Cut algorithm [6] in combination with an energy function inspired by the Boykov function [7]. A set of seeds for the alleged object and the background must be chosen in order to perform the segmentation with the MinCut algorithm; in this work we compared different strategies, some of these exploit the information from the MNOD segmentation masks to better identify the seeds. It is possible to obtain different segmentations for each ROI just by running the MinCut algorithm with different seed extraction strategies. The segmentations obtained from each ROI are processed by a neural model that predicts, for each of them, the degree of membership to each class. Consequently, after a specific ranking algorithm, the class for the object surrounded by each ROI, and the segment that best represents the separation between the object and its background is chosen. The algorithm we devised exploits the information from the segmentation masks, produced by the MNOD model, to limit the number of segments and ROIs that are generated, and then performs a classification of each object.

Section 2 briefly presents the MNOD model on which this work is based, while section 3 presents the multiclass extension proposed in this paper.

2   The Existing Method: Multi-Net Object Detection

The MNOD is a Multi-Net System [8] which consists of an ensemble of supervised neural networks able to detect an object in a cognitive manner, locating the object through the use of a segmentation process. The MNOD is tolerant to many of the problems that usually afflict the images in a real scenario and also has a high generalization ability and good robustness [2].

The MNOD can be represented by a directed acyclic graph, composed by multiple source nodes (which have no incoming edges) used as feature extractors, internal nodes that aggregate the output of other nodes and a single terminal node (which has no outgoing edges) that provides the final segmentation map. Each node n is properly configured with its own parameters $\mathbf{P}$ and acts like an independent module $C_{\mathbf{P}}^{n}$ providing a segmentation map as result. The process starts from the source nodes, which apply operators and filters on the input images in order to generate feature-images which enhance the peculiarities of the input data. Internal nodes process the maps provided by other input nodes firstly

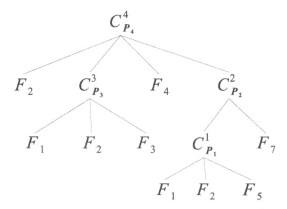

Fig. 1. Generic structure of the MNOD algorithm. The nodes C_P^n represent the supervised neural models which receive their input directly from the source nodes F_i and/or other internal nodes C_P^m.

resizing them according to the parameter I_S. After the resize, internal nodes generate the pattern vectors for the neural network using pixel values that fall within a sliding window of size W_S and gives as output a map image where each pixel has an intensity value proportional to the probability that it belongs to the object. In the present work, each internal node consists of a feed-forward Multi-Layer Perceptron (MLP) trained using the Resilient Backpropagation learning algorithm proposed by Riedmiller and Braun [9].

The particular aspect of this model lies in the connections among the nodes, since the links among the nodes in the structure define the flow of the image segmentation process that passes through the whole structure from the source nodes to the terminal node containing the final segmentation. Source nodes extract different information from the input images; as an example we may use color channels and brightness (Br) for the luminance of a visual target. As suggested in [10], information on edges were extracted using 1-D $[-1, 0, 1]$ masks at $\sigma = 0$, obtaining the two source nodes: Horizontal Edges (HE_s) and Vertical Edges (VE_s), where s is the scale of the input image used as a resize factor to compute the feature.

The MNOD configuration is chosen considering which nodes in which layer gave better results using a trial and error procedure. In particular, we use the following strategy to optimize the results: nodes belonging to the first layer are used to find areas of interest, and nodes of subsequent layers are used to eliminate the false positives and to confirm the true positives.

3   The Proposed Method: Multi-Net Object Detection with Object Classification

One characteristic of the MNOD model presented in the previous section is the possibility to change the node type for each C_P^n. The only requirement is that

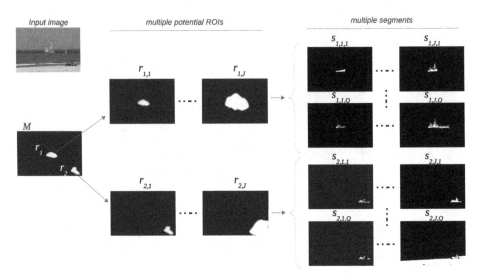

Fig. 2. An example of the segments $s_{k,j,q}$ generated from an input image and an output MNOD map M

every node must implement a common interface. In this work we exploited this feature by integrating a new node $\hat{C}_P^n$ able to classify the potential input ROI into different classes of objects.

More formally, let $\mathbb{C} = \{c_1, \ldots, c_N\}$ be the set of object classes for the classification problem; a node $\hat{C}_P^n$ generates as output a set $\mathbb{I} = \{i_1, \ldots, i_{N+1}\}$ of images, such that i_k is the segmentation mask of all the objects belonging to the class c_k or the background. In order to be able to perform this task, a node $\hat{C}_P^n$ must first generate a set of segmentations of potential objects of interest and then choose which one of these represents the best segmentation. After that, it must select an approriate class for each potential object of interest.

These two steps are described in detail in the following sections 3.1 and 3.2.

3.1 Multiple Segments Creation

A constraint of the proposed strategy imposes only one node of type $\hat{C}_P^n$ as a terminal node that receives as input a single node of type C_P^n. This constraint is adopted in order to simplify the algorithm, but it is not mandatory.

Starting from a segmentation map M, representing the input of the node $\hat{C}_P^n$, the algorithm first identifies all the ROI's $R = \{r_1, \ldots, r_K\}$ for a given input image I. Each r_k corresponds to a connected region identified as in the example showed in figure 2. If the child node worked properly, every r_k contains exactly one of all the objects of interest contained in I. However, the identified r_k may only contain partially the object; this situation can lead to a non-optimal segmentation of the object itself. To maximize the probability that a r_k entirely encloses the object of interest, a set of new ROI's $r_{k,j}$ were generated dilating

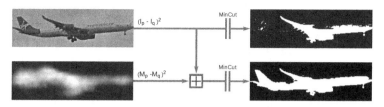

Fig. 3. Comparison between the segmentation results obtained using the energy function proposed by Boykov and our modified version, obtained by introducing an additional energy term that exploits the information contained into the MNOD map M

r_k J times with a mathematical morphology dilation operator, obtaining a set $R_{MP} = \{r_{1,1} \ldots, r_{K,J}\}$ of multiple potential ROI's. For each $r_{k,j}$ we finally generated a set of segments $S = \{s_{k,j,1}, \ldots, s_{k,j,Q}\}$ using a modified version of the Boykov and Komolgorov *min-cut* minimization algorithm [11]. In details, Q corresponds to the number of different strategies, used to initialize the seeds of the *min-cut* algorithm. The same strategy can be applied more than one time, using different parameters.

We modified the energy function proposed by Boykov by adding a new term $T_M = (M_p - M_q)^2$, derived from the segmentation map M, as defined in (1). The use of this additional term helps to ensure good quality segmentations even if the object to be segmented has a color similar to that of the background which surrounds it, as in the example shown in figure 3.

$$V_{p,q} \propto exp\left(-\frac{(I_p - I_q)^2 + (M_p - M_q)^2}{2\sigma^2} \right) \frac{1}{dist(p,q)} \tag{1}$$

To segment each $r_{k,j}$ using the *min-cut* algorithm, a node $\hat{C}_{\mathbf{P}}^n$ defines a set of *seed* pixels P_{obj} belonging to the potential object, and a set of seed pixels P_{bg} belonging to the background, based on information derived from the bounding box of the same ROI $r_{k,j}$. In this paper, we used $Q = 5$ different strategies for the creation of all the seeds (see example in figure 4). In this way we can create a set of possible segments $\{s_{k,j,1}, \ldots, s_{k,j,Q}\}$ for each $r_{k,j}$, as summarized in the example shown in figure 2. Below are briefly presented all the algorithms used to create the seeds. Many of these algorithms exploit the soft segmentation map M, identifying the minimum intensity value of p_{min}, and the maximum p_{max}, in order to extract the seeds.

Random. This strategy extracts a random set of seed P_{obj} from a neighborhood of p_{max} and the set of seed P_{bg} from a neighborhood of p_{min}. The cardinality of the two sets of seeds is a parameter and is such that $|P_{obj}| = |P_{bg}|$.

MaxMin. Similar to the random strategy, but it selects as seeds P_{obj} all the pixels that have value exactly equal to p_{max}, while selects as seeds P_{bg} all the pixels that have value exactly equal to p_{min}.

Distance. The selection of seeds P_{obj} is exactly equal to the random strategy, while to select the seeds P_{bg} this strategy also maximizes the Euclidean distance from the centroid of the P_{obj} set.

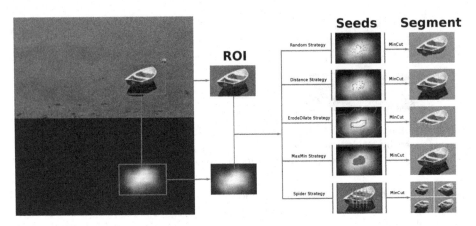

Fig. 4. Example of the segments produced by the *min-cut* algorithm, varying the algorithm that generates the set of seeds

ErodeDilate. This strategy applies the morphological operator erosion to the ROI at $r_{k,j}$ and takes all the edge pixels as belonging to the set of seeds P_{obj}, while to obtain the seeds P_{bg} applies the morphological operator dilation still to the ROI $r_{k,j}$.

Spider. This strategy operates in a completely different way and is inspired by the seed selection technique used in [12]. The pixels that are located near the edges of the ROI $r_{k,j}$ are marked as belonging to P_{bg}; the internal portion of $r_{k,j}$ is divided into N rectangles $\{f_1, \ldots, f_N\}$ of equal size, for each of which a set of seeds P_{obj} is extracted. In this way we obtain a segmentation for each rectangle f_n.

3.2   Classification of the Winner Segment

The final step of a node $\hat{C}_{\mathbf{P}}^n$ is to decide what is the segment that best replaces a ROI r_k, and what is the class of this segment. To do this, a node $\hat{C}_{\mathbf{P}}^n$ uses a trained MLP neural network, which receives as input a set of features extracted from each segment $s_{k,j,q}$, and generates as output N values $\langle o_{1,j,q}, \ldots, o_{N,j,q} \rangle$. Each value $o_{n,j,q}$ is the degree of membership of the segment $s_{k,j,q}$ to the class $c_n \in \mathbb{C}$. An input pattern for this neural model is constructed by combining the following descriptors for each segment $s_{k,j,q}$ processed: PHOG [13], Gray Level Histogram, Area and Perimeter of the segment. These features were chosen after a careful analysis, documented in the section 4.

Starting from the set of predictions made by the neural model on the set of segments $\{s_{k,j,1}, \ldots, s_{k,j,Q}\}$, we determine the class of the object under consideration (eq. 2) as the index w of the first component with maximum value, adding up all the Q predictions:

$$w = \arg\max_{n=1,\ldots,N} \sum_{j=1}^{J} \sum_{q=1}^{Q} o_{n,j,q} \tag{2}$$

After identifying a proper class for the ROI r_k, we choose the best segment $s_{k,j,q}$ $\forall j, q$. In order to do that, all the candidate segments are submitted as input to the trained neural model. The segment $s_{k,j,q}$ is the winner if it generates an output component $o_{w,j,q}$ having a maximum value when compared with all other $J \times Q$ values.

4   Experiments

In this section we evaluate two aspects of our algorithm: the best combination of features for object classification in a real context, and the analysis of the proposed $\hat{C}_P^n$ node when integrated in the existing MNOD model. To assess the performance of the classification and the segmentation phases, we employed the following evaluation metric:

$$OA = \frac{TP}{TP + FP + FN} \tag{3}$$

where the true positives (TP), false positives (FP) and false negative (FN) are computed considering the whole image (obtaining the overall classification accuracy OA_c), or considering the individual pixels (obtaining the overall segmentation accuracy OA_s). For all the experiments we used a computer having the following configuration: a single C# thread, on an Intel®Core ™i5 CPU at 2.67GHz.

In the first experiment the most successful features for image classification were analyzed independently: PHOG [13], PHOW [14], Geometric Blur [15] and Visual Self-Similarity [16]. We employed the Drezzy-46[1] dataset, composed by images representing clothing in the context of online shopping, firstly presented in [17]. The number of images per class varies from a minimum of 52, to a maximum of 227. The dataset contains 4841 images, divided into 46 classes, having a resolution equals to 200x200. Our goal was to determine the most effective feature for the classification of objects belonging to the Drezzy-46 dataset. Varying the features parameters we obtained the highest OA_c using the PHOG feature computed on the segments $s_{k,j,q}$.

The next step attempts to combine the PHOG feature with other simpler features, in order to try to increase the classification accuracy. In particular, we tried to combine the segment area (A), segment perimeter (P) and gray level histogram (GLH) features. Table 1 shows that the use of the feature PHOG in combination with other features allows to achieve a slight increase in terms of accuracy of classification without increasing the computational time. This result allowed us to determine the final configuration: PHOG with 3 layers and 6 bins for each histogram, P, A and GL.

In the second experiment conducted, we analyzed the segmentation accuracy OA_s, using two datasets of manually segmented real images. As the first dataset we employed the DrezzyDataset[2], proposed in [18], considering only three classes:

---

[1] The dataset is available online at: http://artelab.dicom.uninsubria.it/download/
[2] The dataset is available online as: http://artelab.dicom.uninsubria.it/download/

Table 1. OA_c computed on the dataset Drezzy-46, using different combinations of features for the patterns creation

Feature	Test OA_c	Time (ms)
PHOG 2L/06B	0,79	5,0
PHOG 3L/06B	0,82	5,4
PHOG 3L/15B	0,82	6,8
PHOG 2L/06B ∪ A ∪ P ∪ GLH	0,79	4,5
PHOG 3L/06B ∪ A ∪ P ∪ GLH	**0,83**	5,7
PHOG 3L/15B ∪ A ∪ P ∪ GLH	0,81	7,3
PHOG 2L/06B ∪ P ∪ GLH	0,79	5,1
PHOG 3L/06B ∪ P ∪ GLH	0,82	5,6
PHOG 3L/15B ∪ P ∪ GLH	0,80	7,3
PHOG 2L/06B ∪ GLH	0,76	5,0
PHOG 3L/06B ∪ GLH	0,81	5,5
PHOG 3L/15B ∪ GLH	0,82	7,2
A ∪ P ∪ GLH	0,64	0,1

Table 2. OA_s computed on the DrezzyDataset. The same model was tested, using the same configuration, with/without R_{MP} regions and with/without the modified energy term T_M.

Category	without R_{MP}	with R_{MP}	
		without T_M	with T_M
Hat	58,30	60,76	60,91
Shoe	76,66	80,00	80,24
Tie	83,59	87,45	88,21

Hat, Shoe and Tie. To make possible a comparison with other methods, we also employed the VOC2011 [19] dataset. Using the set of R_{MP}, and comparing the results obtained on the DrezzyDataset with the version that does not use this set, we obtained the results presented in table 2. It is possible to observe how the use of such set allows to obtain an increase in terms of OA_s. Obviously the R_{MP} extraction causes an increase in terms of time required to perform the segmentation and classification of a single object; this increase depends on the parameter J. The results presented in table 2 were obtained using only the MaxMin strategy, setting the number of R_{MP} regions to $J = 5$. In the same table we reported the results obtained using the modified Boykov energy function (eq. 1). The use of the term T_M leads to a slight increase in terms of OA_s, while the time required to perform the classification and segmentation of a single object remains unchanged.

Table 3 shows the time required by the final model to perform a classification and segmentation of a single image, varying the image resolution W_s, and using three different configurations of seeds extraction algorithms. Using

Table 3. Time needed to train the model showed in figure 5, and to both classify and segment a single object, varying the size of the image processed by the node $\hat{C}_{\mathbf{P}}^n$, and the seed extraction strategy. The presented values were obtained by averaging on 10 independent executions of the proposed model.

W_s	MaxMin		3x Random(20)		5x Random(20)	
	Train (sec)	Test per Obj. (sec)	Train (sec)	Test per Obj. (sec)	Train (sec)	Test per Obj. (sec)
100x100	14,41	0,48	33,16	0,64	58,46	0,92
150x150	26,16	0,88	68,35	1,13	112,56	1,78
200x200	38,93	1,55	118,43	2,19	207,70	2,96
250x250	61,10	2,18	163,30	4,43	302,10	5,78
300x300	86,99	2,82	227,09	7,08	523,04	8,67

Table 4. OA_s obtained on the dataset DrezzyDataset, using the overall strategy

Category	Test OA_s
Background	90,56
Hat	64,81
Shoe	68,13
Tie	74,09
Overall	74,02

Table 5. Comparison between the OA_s obtained on a subset of classes of the VOC2011 dataset, using the existing MNOD algorithm and the proposed model

Category	Multi Class using $\hat{C}_{\mathbf{P}}^n$ node	Single Class without $\hat{C}_{\mathbf{P}}^n$ node	using $\hat{C}_{\mathbf{P}}^n$ node
Background	83,10	//	//
Aeroplane	23,13	43,12	36,91
Bicycle	3,56	9,58	12,14
Boat	15,66	24,19	21,67
Bus	25,01	58,51	48,72
Car	20,55	33,87	33,24
Motorbike	23,80	50,51	36,02
Train	21,39	48,48	35,92
Overall	27,02	38,32	32,08

$W_s = 200 \times 200$ (typical size of the images belonging to the DrezzyDataset), the proposed model needs approximately 2.9 *sec* to perform the classification and segmentation of an image, using 5 times the Random seed extraction strategy.

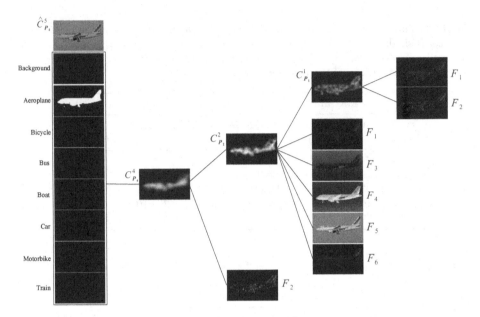

Fig. 5. Topology of a MNOD model using a node $\hat{C}_\mathbf{P}^n$; this configuration is used to classify and to segment the objects belonging to a subset of classes of the VOC2011 dataset

The model is therefore very efficient when compared with other algorithms presented in the literature that perform at the same time the classification and segmentation.

Finally, we evaluated the overall segmentation and classification strategy on the DrezzyDataset. These results are very encouraging and are presented in table 4; however, our results are inferior to those obtained using the existing model in [18], mainly because we need to perform both segmentation and classification of each object of interest. In table 5 we present the results obtained in the segmentation and classification of the images belonging to some classes of the Pascal VOC2011 dataset. The classification of segments with the proposed model leads to a considerable decrease of accuracy, caused mainly by the features used in the classification stage; infact, they reveal to be not significant for the objects belonging to the classes of the dataset VOC2011.

5   Conclusions

The proposed model $\hat{C}_\mathbf{P}^n$ can be integrated into the existing MNOD algorithm in order to realize a multiclass segmentation. The choice of using multiple potential ROIs, and the introduction of the MNOD energy term, leaded to an increase in overall accuracy in both the classification and segmentation phases if compared to the base model. A MNOD network that makes use of the node $\hat{C}_\mathbf{P}^n$ can be trained in a reasonable time; setting up an optimal configuration, the evaluation of a single image takes less than three seconds. The fact that the developed features strongly depend on the application context is a major issue to take into consideration.

The use of the *min-cut* algorithm with our modified Boykov energy function permits to exploit the information from the soft segmentation masks produced by the sliding window nodes. Thus, the choice of setting the position of the seeds starting from the available detection mask led to a good solution as it allows to considerably reduce the number of segments to be generated. However, we could not define a stable and optimal seed extraction strategy, which always produces the same set of seeds starting from the same detection mask.

The proposed model inspires interesting future developments such as: development of new features, more general than the proposed ones; development of a stable and optimal seed extraction strategy; employment of more than one classification node in the MNOD network, both in cascade or in parallel.

References

1. Felzenszwalb, P.F., Mcallester, D.A., Ramanan, D.: A discriminatively trained, multiscale, deformable part model. In: Proceedings of International Conference of Computer Vision and Pattern Recognition (2008)
2. Gallo, I., Nodari, A.: Learning object detection using multiple neural netwoks. In: Proceedings of International Joint Conference on Computer Vision, Imaging and Computer Graphics Theory and Applications. INSTICC Press (2011)
3. Li, F., Carreira, J., Sminchisescu, C.: Object recognition as ranking holistic figure-ground hypotheses. In: Proceedings of International Conference on Computer Vision and Pattern Recognition, pp. 1712–1719. IEEE (2010)
4. Carreira, J., Sminchisescu, C.: Constrained parametric min-cuts for automatic object segmentation. In: Proceedings of International Conference on Computer Vision and Pattern Recognition (2010)
5. Serre, T., Wolf, L., Bileschi, S., Riesenhuber, M., Poggio, T.: Robust object recognition with cortex-like mechanisms. IEEE Transactions on Pattern Analysis and Machine Intelligence 29, 411–426 (2007)
6. Greig, D.M., Porteous, B.T., Seheult, A.H.: Exact maximum a posteriori estimation for binary images. Journal of the Royal Statistical Society, 271–279 (1989)
7. Boykov, Y.Y., Jolly, M.P.: Interactive graph cuts for optimal boundary & region segmentation of objects in n-d images. In: Proceedings of IEEE International Conference on Computer Vision, vol. 1 (2001)
8. Sharkey, A.J.: Multi-Net Systems. In: Combining Artificial Neural Nets: Ensemble and Modular Multi-Net Systems. Springer (1999)
9. Riedmiller, M., Braun, H.: A direct adaptive method for faster backpropagation learning: the RPROP algorithm. In: Proceedings of IEEE International Conference on Neural Networks, pp. 586–591 (1993)
10. Dalal, N., Triggs, B.: Histograms of oriented gradients for human detection. In: Proceedings of International Conference on Computer Vision and Pattern Recognition, pp. 886–893 (2005)
11. Boykov, Y., Kolmogorov, V.: An experimental comparison of min-cut/max-flow algorithms for energy minimization in vision. IEEE Transactions on Pattern Analysis and Machine Intelligence 26, 1124–1137 (2004)

12. Carreira, J., Sminchisescu, C.: CPMC: Automatic object segmentation using constrained parametric min-cuts. IEEE Transactions on Pattern Analysis and Machine Intelligence (2012)
13. Bosch, A., Zisserman, A., Munoz, X.: Representing shape with a spatial pyramid kernel. In: Proceedings of the 6th ACM International Conference on Image and Video Retrieval, CIVR 2007, pp. 401–408. ACM (2007)
14. Lowe, D.G.: Object recognition from local scale-invariant features. In: Proceedings of the International Conference on Computer Vision. IEEE Computer Society (1999)
15. Berg, A.C., Malik, J.: Geometric blur for template matching. In: Proceedings of International Conference on Computer Vision and Pattern Recognition, vol. 1(C), pp. 607–614 (2001)
16. Shechtman, E., Irani, M.: Matching local self-similarities across images and videos. In: Proceedings of International Conference on Computer Vision and Pattern Recognition (2007)
17. Nodari, A., Ghiringhelli, M., Zamberletti, A., Albertini, S., Vanetti, M., Gallo, I.: A mobile visual search application for content based image retrieval in the fashion domain. In: Workshop on Content-Based Multimedia Indexing (2012)
18. Albertini, S., Gallo, I., Vanetti, M., Nodari, A.: Learning object segmentation using a multi network segment classification approach. In: Proceedings of International Conference on Computer Vision Theory and Applications (2012)
19. Everingham, M., Van Gool, L., Williams, C.K.I., Winn, J., Zisserman, A.: The pascal visual object classes challenge (VOC) 2007-2011 results

A Decision Support System for the Prediction of the Trabecular Fracture Zone

Vasileios Korfiatis, Simone Tassani, and George K. Matsopoulos

Institute of Communication and Computer System,
9 Iroon Polytechniou Street, 157 80, Zografou, Athens, Greece
tassani.simone@gmail.com

Abstract. Prediction of trabecular fracture zone is a very important element for assessing the fracture risk in patients. The assumption that failure always occurs in local bands, the so called 'fracture zones', with the remaining regions of the structure largely unaffected has been visually verified. Researchers agreed that the identification of the weakest link in the trabecular framework can lead to the prediction of the fracture zone and consequently of the failure event. In this paper, a decision support system (DSS) is proposed for the automatic identification of fracture zone. Initially, an automatic methodological approach based on image processing is applied for the automatic identification of trabecular bone fracture zone in micro-CT datasets, after mechanical testing. Then, a local analysis of the whole specimen is performed on order to compare the structure (Volumes of Interest -VOI) of the broken region to the unbroken one. As a result, for every VOI, 29 morphometrical parameters were computed and used as initial features to the proposed DSS. The DSS comprises of two main modules: the feature selection module and the classifier. The feature selection module is used for reducing the initial size of the input features' subset (29 features) and for keeping the most informative features in order to increase the classification's module performance. To this end, the Sequential Floating Forward Selection (SFFS) algorithm with Fuzzy C-Means evaluation criterion was implemented. For the classification, several classification algorithms including the Multilayer Perceptron (MLP), the Support Vector Machines (SVM), the Naïve Bayesian (NB), the k-Nearest Neighbor (KNN) and the k-Means (KM) have been used. Comparing the performance of these classification algorithms, the SFFS-SVM scheme provided the best performance scoring 98% in terms of overall classification accuracy.

Keywords: Trabecular Fracture Zone, Decision Support Systems, Machine Learning, Feature Selection, Sequential Floating Forward Selection (SFFS), Multilayer Perceptron (MLP), Support Vector Machines (SVM), Naïve Bayesian (NB), k-Nearest Neighbor (KNN), k-Means (KM).

1   Introduction

Prediction of trabecular fracture zone is a wide challenge not yet solved. The cost of fracture treatments related to fracture event are extremely high in Europe and all over the world. The cost related to years of healthy life lost due to disability is also high.

N. Mana, F. Schwenker, and E. Trentin (Eds.): ANNPR 2012, LNAI 7477, pp. 163–174, 2012.
© Springer-Verlag Berlin Heidelberg 2012

For this reason the right prediction of the fracture zone is a mandatory step for the study and assessment of the fracture risk in patients.

Some in-vitro studies pointed out the importance of local variation of the framework for the global behavior of the whole trabecular structure [1-3]. Researchers agree that the identification of the weakest link in the trabecular framework can lead to the prediction of the fracture zone, first, and of the failure event in general. In a previous study the visual identification of the trabecular fracture zone was used in order to analyze the morphometrical characteristics of the fracture zone and use them to predict the fracture [2]. This approach led to a classification error of about 20%. Nonetheless, the visual identification of the fracture zone limited the study in time and accuracy. Recently an automatic scheme for the identification of the trabecular fracture zone was presented and validated [4]. The automatic tool allowed performing a fast and accurate three-dimensional identification of the broken region. Preliminary results on the morphometry of trabecular fracture were already presented [5]. Authors underline the morphometrical variability of the trabecular structure. The trabecular broken region was found to be a very small part of the whole specimen, and with statistically significant different morphology. Nonetheless, it was not possible to predict the trabecular fracture zone by means of standard statistical tools based on the analysis of parametrical distributions of mean values and standard deviations.

Recent advances in medicine have shown that artificial intelligence (AI) techniques can be used towards the development and application of decision support systems to aid diagnosis [6]. Learning machines have already shown in various fields the capability to recognize different patterns starting from the analysis of specific features [7-11]. Nonetheless, the application of these techniques in the study of trabecular structure was poorly investigated [12], and, to the authors' knowledge, was never applied to the prediction of the trabecular fracture zone using micro-CT images.

The purpose of this paper is to propose a decision support system (DSS) for the prediction of the trabecular fracture zone using the information of structural differences related to the fracture region. The feature selection module is used for reducing the initial size of the input fea-tures' subset (29 features) and is comprised by the Sequential Floating Forward Selection (SFFS) algorithm with the Fuzzy C-Means evaluation criterion. Many traditional and well-researched classifiers were implemented and compared during the experimental procedure, including the Multilayer Perceptron (MLP), the Support Vector Machines (SVM), the Naïve Bayesian (NB), the k-Nearest Neighbor (KNN) and the k-Means (KM). These classified each pattern into two distinct classes: the non-fractured and the fractured. Additionally in order to improve the performance of the classification algorithms and reduce the size of the input features, a feature selection (FS) pre-processing stage that used the Sequential Floating Forward Selection (SFFS) algorithm was implemented. The experimental results show that the SVM algorithm using the SFFS feature selector provides almost perfect performance.

2   Materials and Methods

2.1   Data Acquisition

One cylindrical trabecular bone specimen (10mm diameter, 24mm height) was extracted from a human femoral condyle of a donor without skeletal disorders during the LHDL (IST-2004-026932) European Project [1]. The specimen was scanned at a pixel size of 19.5μm by means of a micro-CT (model Skyscan 1072, Skyscan, Kontich, Belgium) and mechanically tested (model Mini bionix 858, MTS Systems Corp., Minneapolis, MN, USA) following a previously published protocol [2, 3,13]. After the mechanical compression the specimen was scanned again in micro-CT obtaining two datasets: pre and post-failure.

2.2   Identification of the Fracture and Non-fracture Zones

An automatic registration scheme for the identification of the full 3D broken region was applied [4, 5]. The used method is a surface-based registration technique which involves the determination and matching of common surfaces of the two sets by the minimization of a distance measure [14]. The method was applied on the pre- and post-failure datasets every specimen. The geometrical transformation employed was the rigid transformation model [15]. The 3D automatic registration method was applied as previously described in literature [4, 5]. Every trabecula was classified as broken or not, based on an overlap criterion described and validated in literature [4]: where a trabecula of the after compression dataset overlap more than 30% with the same trabecula of the before compression dataset all the pixels of that trabecular were classified as unbroken, otherwise they were classified as broken.

2.3   Local Analysis

Preliminary studies already showed the extremely local nature of the trabecular fracture [5, 16]. Therefore, a local analysis of the whole specimen was performed on order to compare the structure of the broken region to the unbroken one. The whole unbroken dataset constitutes of a stack of 469x469x991 voxels and it was divided in 1134 cubic volume of interests (VOIs). Every VOI was 93x93x93 voxels and they had an overlap of 46 voxel in all the three dimensions (2D application is shown in Fig. 1).

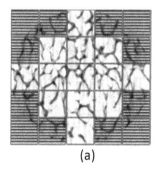 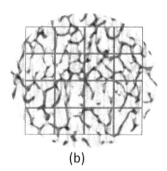

(a)   (b)

Fig. 1. Two dimensional projection of the VOIs selection is shown. a) The specimen was split in cubes. VOIs heaving less than 80% of the volume covered by the cylindrical shape of the specimen were not included in the analysis (shadowed VOIs). b) Translation of the VOIs in x and y is shown.

Due to the cylindrical shape of the specimen, not all the obtained VOIs were suitable for the morphometric analysis. Only VOIs having at least 80% of the volume covered by the cylindrical shape of the specimen were selected (Fig. 1a). The morphometric analysis was finally performed only on the actual tissue volume, defined by the cylindrical geometry of the specimen (i.e. the white part in four corner volumes in Fig. 1b was not included in the tissue study).

2.4   Feature Description

For every VOI, 29 morphometrical parameters were computed. These include 1.tissue volume (TV), 2.bone volume (BV), 3.bone surface (BS), 4.bone volume fraction (BV/TV), 5.bone surface to volume ratio (BS/BV), 6.bone surface density (BS/TV), 7.direct trabecular thickness (Tb.Th*), 8.direct trabecular separation (Tb.Sp*), 9.structure model index (SMI), 10.connectivity density (CD), 11-13.the eigenvalues (E1, E2, and E3) and 14-22.eigenvectors' coordinates of the fabric tensor, 23-25.their normalizations (H1, H2, and H3), 26.the off-axis angle, 27-28.two degrees of anisotropy (DA_1 and DA_2) and 29.the ellipsoid correlation coefficient (CC). The normalized eigenvalues were computed using the normalization proposed by Turner et al. [17], obtained from the three principal directions of the ellipsoid. The above parameters were used as input features for the proposed DSS and were calculated for the VOIs of each specimen using the software "CtAnalyzer" (Skyscan, Kontich, Belgium).

Tissue Volume (TV) is the total volume of every volume-of-interest (VOI). The 3D volume measurement is based on the marching cubes volume model of the VOI. Bone Volume (BV) is the total volume of binarized objects of the foreground within the VOI or in other words the sum of voxels marked as bone. Bone Surface (BS) is the surface area of all the solid objects within the VOI, measured in 3D (Marching cubes method). The calculation of the BV/TV, BS/BV and BS/TV was also used as features.

The Tb.Th*, based on the estimation of volume-based local thickness independently of an assumed structure type, was calculated using the sphere-fitting method [18]. Tb.Sp* is defined as the diameter of the marrow cavities, and was calculated using the same technique described for Tb.Th*[12]. The SMI is a topological index, giving an estimation of the characteristic form in terms of plates and rods composing the 3D structure. For ideal plates and rods, the index gives respectively the values 0 and 3, whereas for a mixed structure the SMI-index lies in between 0 and 3 [19]. The connectivity density is a parameter to measure the degree of multiple connections, and hence reports the maximal number of branches that can be broken in a network before the structure is separated into two parts [20]. The term fabric was introduced in bone mechanics as a description of the anisotropy of a material's microstructure [21]. The fabric tensor was defined as a positive second rank tensor which quantitatively describes fabric [21, 22].

The eigenvalues of the fabric tensor (E_i) were obtained using the mean intercept length (MIL) technique [23, 24]. Using this technique a grid of parallel line is superimposed to the trabecular structure. The average distance between interceptions of the grid with the trabecular structure is computed and plotted in 3D. If the structure is orthotropic (three axis of symmetry) the distribution can be approximated to an ellipsoid. The values of E_1, E_2, and E_3 represent the principal axes in three dimensions (3D) of the ellipsoid. These values give information about how distant the trabeculae are in the three dimensions, and therefore how oriented the framework is in every direction. These eigenvalues and the consequent eigenvectors' coordinates (which are the 3x3 fabric tensor matrix's values) give twelve features in total. E_i were shown to be dependent from other structural parameters (e.g. the porosity of the structure) [25], therefore the normalization of the eigenvalues was also used. Normalized eigenvalues (H_i) were obtained from the three principal directions of the ellipsoid, using the normalization proposed by Turner et al. [17]:

$$H_1 \pm H_2 \pm H_3 = 1 \ (1)$$

Degree of anisotropy (DA) is a measurement of the anisotropy of the 3D object. Anisotropy is a measure of 3D asymmetry along a particular directional axis. Apart from percent volume, DA and the general stereology parameters of trabecular bone are probably the most important determinants of mechanical strength [22]. The two DAs analyzed are the ones suggested in the literature [26, 27] and were computed as a ration between the three principal MILs:

$$DA_1 = MIL_1/MIL_3 \ (2)$$

$$DA_2 = MIL_2/MIL_3 \ (3)$$

The off axis angle was defined as the difference between the main trabecular direction of the analyzed VOI, identified by the direction of the principal eigenvector, and the load direction. Finally, the CC's value corresponds to the fitting of the ideal ellipsoid to the experimental one.

2.5   Proposed Decision Support System

The proposed decision support system (DSS) is comprised by the following modules (Fig. 2):

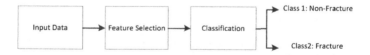

Fig. 2. The proposed DSS

The input data consists of 734 patterns in total. From these 453 belong to the non-fractured class and 281 to the fractured class, providing an approximate ratio of 60/40% non-fractured/fractured patterns. The original data set contained 1134 patterns, which represent the VOIs of the specimen, but its composition was very unbalanced in favor of the non-fractured class. As a result 400 patterns from the non-fractured class were randomly removed, and the rest 734 were fed to the proposed DSS.

The feature selection module is used for reducing the initial size of the input features' subset (29 features) and for keeping the most informative features in order to increase the classification's module performance. To this end, the Sequential Floating Forward Selection (SFFS) algorithm [28] with Fuzzy C-Means [29] evaluation criterion was implemented. This algorithm starts with a feature subset of a certain predetermined size and moves up or down in number in order to find the best composition. It usually converges to a local minimum, thus is considered suboptimal. The feature subset that the algorithm returns generally tries not to deviate much in size from the given starting size. For the experimental procedure the number of epochs was set to 100 and the number of starting features was set to four values, 5, 10, 15 and 20.

In the classification module, five well-established classification algorithms were implemented including the Multilayer Perceptrons (MLP) [30], Support Vector Machines (SVM)[31], Naïve Bayesian (NB)[32], k-nearest neighbor (KNN)[32] and k-means clustering (KM)[33]. The MLP had one hidden layer with 20 neurons and used the linear transfer function for the first stage (input-hidden) and the hyperbolic tangent sigmoid transfer function for the second stage (hidden-output). For its training the error back-propagation technique with the Levenberg-Marquardt optimization was used. The SVM used the Sequential Minimal Optimization (SMO) technique for finding the separating hyperplane and the linear kernel function (dot product) for mapping the training data into kernel space. For the NB the kernel smoothing density estimate was used for modelling the distribution of the data. The kernel used was the normal (Gaussian) kernel and the width was selected automatically from the Matlab 2011b software for each class and feature.The prior probabilities for the classed were specified using the empirical rule which takes into account the relative frequencies of the classes in training. The KNN classifier used the nearest neighbor only (k=1) and the Euclidean distance for classifying the test data. Finally, for the KM the number of centroids was set to k=2 since the problems consists of two classes and the distance measure used was the Squared Euclidean with an online update phase.

For the supervised learning algorithms (MLP, SVM, NB and KNN) the cross-validation training procedure [34] was used in order to provide convenient statistical quality of the results. The data set was randomly broken down into five groups so that each group contained 1/5 of the total given patterns. As a result, the technique was five-fold cross-validation and the training-test ratio was 80/20%. For each of the five repetitions, one of the groups was considered the test set and the other four merged and created the training set. The mean values from all five repetitions formed the final values of the measures for each algorithmic scheme.

For the unsupervised learning KM, there are no training and test sets and all the selected features are fed to the algorithm. The algorithm stars with the two centroids in random locations and then assigns each pattern to the closest centroid according to the selected distance measure, creating the two clusters. It then changes the positions of the two centroids in the feature space to the average location of the selected patterns of each and repeats the above procedure. It finishes when no change in the centroids' coordinates is made. For the evaluation of the results the true labels of the patterns were compared to the labels (clusters) that the KM assigned. Since the class-cluster assignment is interchangeable for each algorithmic run, the assignment that provided the higher value of the evaluation measure was considered valid each time.

2.6   Evaluation Measures

Based on the confusion matrices of each classifier scheme the accuracy, the sensitivity and the specificity were calculated using the following equations:

$$Accuracy = \frac{Number\ of\ Instances\ Labeled\ Correctly}{Total\ number\ of\ Instances} \tag{4}$$

$$Sensitivity = \frac{True\ Positives}{True\ Positives + False\ Negatives} \tag{5}$$

$$Specificity = \frac{True\ Negatives}{True\ Positives + False\ Positives} \tag{6}$$

In simpler terms sensitivity measures how precise is the classifier in predicting the class A, which is the non-fractured in our experiment and the specificity the same for class B, which is the fractured. We also used ROC space representation in order to more accurately assess the quality of the implemented classifiers.

For the experiments an Intel Q9450@2.66 GHz CPU, 4GB DDR2 RAM PC running on Windows 7 OS was used. All machine learning techniques were developed using Matlab 2011b.

3   Results

The best system performance is obtained using the SFFS with 15 features and the SVM algorithm. In that case, the sensitivity 99%, the specificity 95% and the total accuracy is 98%. Fig. 3 is a multiple column graph that contains the visualisations of the percent measurements of accuracy, sensitivity and specificity for the five implemented classifiers. These are the mean values from 5 repetitions. Each measure is represented by a different shade of grey. That information is illustrated simultaneously for all the different feature selection scenarios that were tried out (5, 10, 15 and 20 starting features) and for the scenario that no feature selection was used.

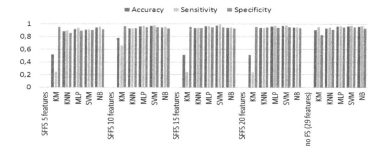

Fig. 3. Experimental results for the various classification scenarios

In Table 1, the confusion matrices of the classification for the best feature scenario (15 features) are provided. These results are the mean values of 5 different algorithm executions. Rows correspond to the real labels and columns to the classifier's labels. Class 1 is the name of the non-fractured class (90 test-patterns for each iteration) and class 2 of the fractured one (56 test-patterns for each iteration). The KM clustering is a non-supervised classification algorithm and as a result the total number of patterns is taken into account. From the Table 1, it is evident that the SVM shows the best performance.

Table 1. Confusion Matrix for SFFS-15 Features Averaged Over 5 Independent Algorithm Executions

Method	Actual class	Predicted class	
		Class 1	Class 2
KM	Class 1	111	342
	Class 2	13	268
KNN	Class 1	86	3.8
	Class 2	4.6	52.4
MLP	Class 1	87.8	2.8
	Class 2	2.8	53.4
SVM	Class 1	90.2	3
	Class 2	0.4	53.2
NB	Class 1	86.2	3.8
	Class 2	4.4	52.4

In Fig.4, the ROC curves of the classifiers for the SFFS with 15 features are shown. In the Y-axis is the True Positive Rate (TPR) which is equal to the Sensitivity and in the X-axis is the False Positive Rate (FPR) which is equal to (1-Specificity). These curves prove again that the SVM is the best classifier followed closely by the MLP and the rest of the supervised classifiers.

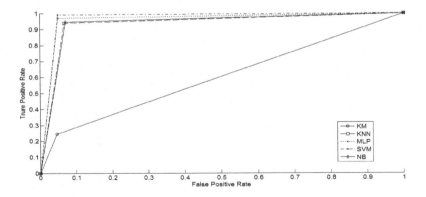

Fig. 4. ROC curves

4   Discussion

Trabecular fracture is a much localized event. This scenario brought a natural imbalance between the numbers of gathered samples that belonged to the non-fractured class and those that belonged to the fractured class, with the first being much greater than the second. The original dataset contained 1134 patterns from which 853 belonged to the non-fractured class and 281 to fractured class, striking a ration of approximately 75-25%. It is common knowledge though that the two most important attributes of a good data set are having a big number of patterns and striking a balanced ratio among the different classes. As a result, 400 patterns that belonged to the non-fractured class were removed, but all the patterns that belonged to the fractured class were kept, producing the final experimental data set that was described earlier. This choice greatly enhanced the quality of the classifiers in every aspect.

The feature selection stage was implemented in an attempt to improve the system's performance and at the same time investigate the impact of the several input features on the classification procedure. The important parameters of the SFFS are the number of starting features (input features' subset size) and the number of epochs, which represents the maximum allowed repetitions in case the algorithm does not converge in a minimum. The experimental results show clearly that the general performance of all the supervised classifiers is slightly affected by the presence of a feature selection system. For example, the SVM's accuracy was 91% for 5 starting features, 96% for 10 starting features, 98% for 15 starting features (best case) and 96% for 20 starting features. The other supervised classifiers showed even smaller deviation as the features' number increased but still could not outperform the SVM. The performance of the KM clustering (unsupervised) showed great variance in terms of the starting feature parameter's value. Its results remain poor though except from the case when no FS is used; then they are acceptable but still inferior from the rest. Considering the features number it seems that the best option is around 15 features. Much lower values of the starting features parameter decrease the performance whereas higher values do not provide any significant improvement. The algorithms always converged within the given number of epochs. Its running time was analogous of the starting features and was about 2 minutes for the 15 starting features case.

In fig.5 the frequency of selection for each input feature is illustrated. Consulting the numbering given in the feature presentation in Chapter 2 the most important features, appear to be the direct Tb.Th, the direct Tb.Sp, the BV and the 'y' component of the 'y' eigenvector , which were selected every time the SFFS algorithm run.

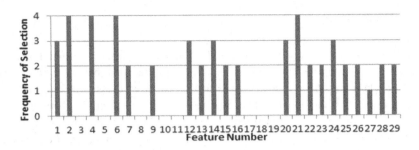

Fig. 5. Frequency of selection for each feature

Finally, it has to be noted that the only FS algorithm that was implemented was the SFFS. As a result it is possible that other FS algorithms may provide better performance and/or indicate other features as important. Nonetheless, it is important to notice that, among the 4 parameters always selected by the SFFS, 3 are related to the amount of tissue and only one is related to the orientations of the structure. However, the meaning of Tb.Th and Tb.Sp analyzed together is related to the number of trabeculae within the analyzed VOI. Therefore, it is suggested that the number of trabeculae, the amount of bone and the orientation of the structure are the most important parameters, leaving out of the discussion the anisotropy of the tissue, the connectivity of the structure and the plate/rod shape of the trabeculae. Nonetheless, even if 734 specimens were analyzed, we must remember that the whole study is related to the analysis of only one physical specimen. A bigger sample size should be analyzed before any reliable physical meaning could be carried out.

The importance of the features seems the most interesting topic though since the performance achieved without the FS stage is generally very good (over 90% accuracy for all implemented schemes). Nonetheless, FS is a mandatory step in order to move to the clinical domain. Micro-CT images of in-vitro specimens allow to perform very accurate analysis and to compute a great number of parameters. However, it is strongly improbably to be able, even in the future, to perform such in deep analysis on clinical images. For this reason it is important to identify the most important parameters and find a way to compute the same parameters in clinics.

Considering the optimal classifier the 15 features scheme will be taken into account. In that case the performance of all the supervised classifiers is very good but the most robust and efficient option for this binary classification problem seems to be the SVM. It performed almost perfectly and needed in general very low amount of training time (about 1 second). Concerning the MLP training time the number of epochs it needed was around 10, a very low number as well. It has to be noted that the unsupervised KM clustering performed worse than all the supervised schemes, mainly because it showed a clear tendency in classifying the majority of the patterns to the fractured class.

References

1. Tassani, S., Ohman, C., Baruffaldi, F., Baleani, M., Viceconti, M.: Volume to density relation in adult human bone tissue. J. Biomech. 44, 103–108 (2011)
2. Perilli, E., Baleani, M., Ohman, C., Fognani, R., Baruffaldi, F., Viceconti, M.: Dependence of mechanical compressive strength on local variations in microarchitecture in cancellous bone of proximal human femur. J. Biomech. 41, 438–446 (2008)
3. Tassani, S., Ohman, C., Baleani, M., Baruffaldi, F., Viceconti, M.: Anisotropy and inhomogeneity of the trabecular structure can describe the mechanical strength of osteoarthritic cancellous bone. J. Biomech. 43, 1160–1166 (2010)
4. Tassani, S., Asvestas, P.A., Matsopoulos, G.K., Baruffaldi, F.: Automatic Identification of Trabecular Bone Fracture. In: Bamidis, P.D., Pallikarakis, N. (eds.) MEDICON 2010. IFMBE Proceedings, vol. 29, pp. 296–299. Springer, Heidelberg (2010)
5. Tassani, S., Demenegas, F., Matsopoulos, G.K.: Local analysis of trabecular bone fracture. In: Conf. Proc. IEEE Eng. Med. Biol. Soc., pp. 7454–7457 (2011)
6. Leroy, G., Chen, H.: Introduction to the special issue on decision support in medicine. Decision Support Systems 43, 1203–1206 (2007)
7. Palaniswami, M.: Computational Intelligence in Gait Research: A Perspective on Current Applications and Future Challenges. IEEE Transactions on Information Technology in Biomedicine 13(5), 687–702 (2009)
8. Mashor, M.Y., Jaafar, H.: Online sequential extreme learning machine for classification of mycobacterium tuberculosis in ziehl-neelsen stained tissue. In: 2012 International Conference on Biomedical Engineering, February 27-28, pp. 139–143 (2012)
9. Pena, E., Martínez, M.A.: Machine Learning Techniques as a Helpful Tool Toward Determination of Plaque Vulnerability. IEEE Transactions on Information Technology in Biomedicine 59(4), 1155–1161 (2012)
10. Madokoro, H., Otani, T., Kadowaki, S.: Experimental studies with a hybrid model of unsupervised neural networks. In: The 2011 International Joint Conference on Neural Networks (IJCNN), July 31-August 5, pp. 1659–1666 (2011)
11. Phlypo, R., Congedo, M.: SVM feature selection for multidimensional EEG data. In: IEEE International Conference on Acoustics, Speech and Signal Processing (ICASSP), May 22-27, pp. 781–784 (2011)
12. Christopher, J.J., Ramakrishnan, S.: Assessment and classification of mechanical strength components of human femur trabecular bone using texture analysis and neural network. J. Med. Syst. 32, 117–122 (2008)
13. Ohman, C., Baleani, M., Perilli, E., Dall'Ara, E., Tassani, S., Baruffaldi, F., Viceconti, M.: Mechanical testing of cancellous bone from the femoral head: experimental errors due to off-axis measurements. J. Biomech. 40, 2426–2433 (2007)
14. Matsopoulos, G.K., Delibasis, K.K., Mouravliansky, N.A., Asvestas, P.A., Nikita, K.S., Kouloulias, V.E., Uzunoglu, N.K.: CT-MRI automatic surface-based registration schemes combining global and local optimization techniques. Technol. Health Care 11, 219–232 (2003)
15. van den Elsen, P.A., Pol, E.J.D., Viergever, M.A.: Medical image matching-a review with classification. IEEE Engineering in Medicine and Biology Magazine 12, 26–39 (1993)
16. Tassani, S., Demenegas, F., Matsopoulos, G.K.: Morphometry of trabecular bone fracture: preliminary study. In: XXIIIrd Congress of the International Society of Biomechanics, July 3-7, p. 69. International Society of Biomechanics, Brussels (2011)
17. Turner, C.H., Cowin, S.C., Rho, J.Y., Ashman, R.B., Rice, J.C.: The fabric dependence of the orthotropic elastic constants of cancellous bone. J. Biomech. 23, 549–561 (1990)

18. Hildebrand, T., Ruegsegger, P.: A new method for the model-independent assessment of thickness in three-dimensional images. Journal of Microscopy 185, 67 (1997)
19. Hildebrand, T., Ruegsegger, P.: Quantification of Bone Microarchitecture with the Structure Model Index. Comput. Methods Biomech. Biomed. Engin. 1, 15–23 (1997)
20. Odgaard, A., Gundersen, H.J.: Quantification of connectivity in cancellous bone, with special emphasis on 3-D reconstructions. Bone 14, 173–182 (1993)
21. Cowin, S.C.: The relationship between the elasticity tensor and the fabric tensor. Mechanics of Materials 4, 137 (1985)
22. Odgaard, A.: Three-dimensional methods for quantification of cancellous bone architecture. Bone 20, 315–328 (1997)
23. Harrigan, T.P., Mann, R.W.: Characterization of microstructural anisotropy in orthotropic materials using a second rank tensor. Journal of Materials Science 19, 761 (1984)
24. Whitehouse, W.J.: The quantitative morphology of anisotropic trabecular bone. J. Microsc. 101, 153–168 (1974)
25. Tassani, S., Particelli, F., Perilli, E., Traina, F., Baruffaldi, F., Viceconti, M.: Dependence of trabecular structure on bone quantity: a comparison between osteoarthritic and non-pathological bone. Clin. Biomech. (Bristol, Avon) 26, 632–639 (2011)
26. Goulet, R.W., Goldstein, S.A., Ciarelli, M.J., Kuhn, J.L., Brown, M.B., Feldkamp, L.A.: The relationship between the structural and orthogonal compressive properties of trabecular bone. J. Biomech. 27, 375–389 (1994)
27. Majumdar, S., Kothari, M., Augat, P., Newitt, D.C., Link, T.M., Lin, J.C., Lang, T., Lu, Y., Genant, H.K.: High-resolution magnetic resonance imaging: three-dimensional trabecular bone architecture and biomechanical properties. Bone 22, 445–454 (1998)
28. Pudil, P., Novovicova, J., Kittler, J.: Floating search methods in feature selection. Pattern Recognition Letters 15, 1119–1125 (1994)
29. Nock, R., Nielsen, F.: On Weighting Clustering. IEEE Trans. on Pattern Analysis and Machine Intelligence 28(8), 1–13 (2006)
30. Rosenblatt, F.: Principles of Neurodynamics: Perceptrons and the Theory of Brain Mechanisms. Spartan Books, Washington, DC (1961)
31. Cortes, C., Vapnik, V.N.: Support-Vector Networks. Machine Learning 20 (1995)
32. Theodoridis, S., Koutroumbas, K.: Pattern Recognition, 4th edn. Elsevier Inc. (2009)
33. Lloyd, S.P.: Least squares quantization in PCM. IEEE Transactions on Information Theory 28(2), 129–137 (1982)
34. Devijver, P.A., Kittler, J.: Pattern Recognition: A Statistical Approach. Prentice-Hall, London (1982)

Teeth/Palate and Interdental Segmentation Using Artificial Neural Networks

Kelwin Fernandez and Carolina Chang

Grupo de Inteligencia Artificial
Universidad Simón Bolívar
Caracas, Venezuela
kelwin@gia.usb, cchang@usb.ve

Abstract. We present a computational system that combines Artificial Neural Networks and other image processing techniques to achieve teeth/palate segmentation and interdental segmentation in palatal view photographs of the upper jaw. We segment the images into teeth and non-teeth regions. We find the palatal arch by adjusting a curve to the teeth region, and further segment teeth from each other. Best results to detect and segment teeth were obtained with Multilayer Perceptrons trained with the error backpropagation algorithm in comparison to Support Vector Machines. Neural Networks reached up to 87.52% accuracy at the palate segmentation task, and 88.82% at the interdental segmentation task. This is an important initial step towards low-cost, automatic identification of infecto-contagious oral diseases that are related to HIV and AIDS.

Keywords: teeth/palate segmentation, multilayer perceptron, support vector machines.

1 Introduction

Thirty-four million people were living with HIV by the end of 2010 according to the World Health Organization [1]. A 93.5% of these people resided in developing countries, and only 47% of elegible patients in this subgroup received the antiretroviral treatment they needed.

Evidence suggests that about 70% of people with HIV have oral diseases, including Leucoplasia Vellosa, Kaposi's Sarcoma and Candidiasis [2]. The diagnosis of these diseases is very important because some of them may indicate the evolution of HIV towards AIDS [3]. Sadly, the number of dental care centers that treat infecto-contagious deceases in developing countries is limited, and in many cases, insufficient. We believe that some oral diseases that indicate the presence of HIV/AIDS can be diagnosed automatically. A computational, low-cost tool for detecting oral infecto-contagious diseases can lead to health care improvement, especially in low- and middle-income countries.

We focus on the problem of teeth/palate segmentation in palatal view photographs of the upper jaw (maxilla) as an initial step towards diagnosing oral

N. Mana, F. Schwenker, and E. Trentin (Eds.): ANNPR 2012, LNAI 7477, pp. 175–185, 2012.

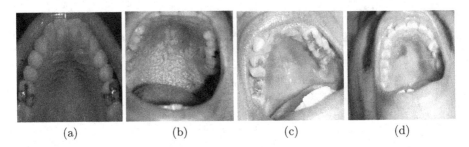

(a)   (b)   (c)   (d)

Fig. 1. (a) Healthy Palate. Oral diseases: (b) Candidiasis, (c) Hepatitis, (d) Kaposi's Sarcoma.

diseases. The palate is an important and large area of the mouth where several diseases can be observed. For example, figure 1 shows cases of patients with oral deseases such as Candidiasis, Hepatitis and Kaposi's Sarcoma. In addition to segmenting the palatal region, we aim to segment each tooth of the upper jaw to give a better description of the teeth region. This information could help refine the palatal region, or detect teeth diseases at future versions of the system.

2   Related Work

Methods for interdental segmentation in dental radiographs for general post-mortem identification are [4] [5]. In [4] a neural network based method is proposed for a postmortem identification system by matching image features extracted from dental radiographs. This system tries to match post-mortem and ante-mortem radiographs of a person. It proposes the use of learnable inherent dental image features for tooth-to-tooth image comparisons. In [5] a dental classification and numbering system to segment, classify, and number teeth in dental bitewing radiographs is proposed. Radiographs are enhanced to isolate teeth to regions of interest using image filters. Once teeth are isolated, a support vector machine classifies each tooth to molar or premolar.

In [13] the problem of lip segmentation in color space is handled using color photographies. The proposed solution set a Gaussian model using the hue and saturation value of each pixel within the lip segment. Then, the memberships of lip and non-lip regions are calculated, and the desirable lip region is obtained based on the memberships.

Radiographs have the advantage of having been taken under common standards, with low-variability machines, but provide less information about diseases compared to palatal view photographs. On the other hand, photographs are highly sensitive to phenomena such as illumination, quality of the image, angle of view, and lack of standards, among other. We address these issues on a variety of palatal view photographs, such as those shown in figure 1.

3   System Overview

The input of the system are palatal view photographs of the upper jaw (maxilla) such as those of figure 1. The system is able to determine if an image holds this constraint. Three stages of the system are described: teeth regions detection, palate segmentation, and interdental segmentation.

The teeth regions detection stage classifies each pixel of the image into teeth/ non-teeth regions based mainly on color features (see section 4). Some pre and post-processing are required to enhance the image quality and the output. We trained Support Vector Machines and Artificial Neural Networks. Experiments show that Neural Networks reached 85.05% accuracy.

At the palate segmentation stage a curve is adjusted to the palatal arch by means of an iterative heuristic process. Outsider teeth regions are discarded after the palatal arch is found. Not only the shape but also the width of the curve is adjusted to cover teeth and leave noise outside. The palate region is considered to be the area inside the closed curve. We achieved 87.52% accuracy at the palate classification stage (see section 5).

At the final stage, Support Vector Machines as well as Artificial Neural Networks were trained to detect the starting and ending points of each tooth. To do so, the curved teeth region obtained in the previous step is flattened, and then inspected through a sliding window. This process is explained in section 6. Results show that Artificial Neural Networks reached 88.82% accuracy at this task.

For the sake of clarity, experiments and results for each stage of the system are presented at the end of its corresponding section. All experiments were carried out on an Intel Core2Duo 2.4 GHz processor, with 4GB of RAM, running Ubuntu Linux 11.04.

4   Teeth Detection

Color is determinant in our teeth segmentation method. A color enhancement filter improves color differentiation and normalizes the color range of images. Frequently, color images are in RGB format. When operators such as histogram equalization and contrast adjustment are applied over each component separately (red, green and blue), new undesirable colors can show up [6]. Hence, we transform original RGB images to HSV format (hue, saturation, value).

Xiao&Ohya proposed a contrast enhancement filter for color images [6]. They handle the problem of invalid new colors in the resulting image by keeping the Hue component unchanged, where the color information lays. The contrast in the V component (luminance value) is enhanced and finally, the saturation component is improved by histogram equalization.

We propose a filter based on this method. Our variation has the same approach in the Hue and Value components. However, in our variation, the Saturation component distribution is shifted such that the mean saturation is the same in all images. This variation tries to keep the difference in the saturation of each pixel while brings a saturation value suitable for distinguishing colors.

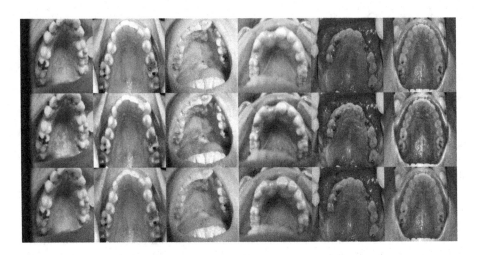

Fig. 2. Image preprocessing. Top: Original images. Middle: Color Enhanced images with Xiao&Ohya's filter. Bottom: Color Enhanced images with the proposed filter.

Finally, in both filters, images are restored to RGB format. These filters are useful in opaque images, where colors are almost indistinguishable. Figure 2 shows the result of applying the filters over six images.

We trained Support Vector Machines and Multilayer perceptrons to classify pixels of the images into teeth/non-teeth regions. SVM kernels include linear, polynomial and RBF kernels. Each SVM was trained varying the parameters using a logarithmic traversing over the search space. The training algorithm of the Multilayer Perceptrons is the classical random sequential back-propagation with a standard sigmoid as the activation function [11]. Neural Networks topologies had one hidden layer, which had from 5 up to 100 hidden neurons.

Experimental results are shown in section 4.1.

A post processing filter is applied to erase small blobs of the output image [10]. A blob is defined as a contiguous set of pixels of similar color. Small blobs are treated as noise. Figure 3 shows some examples of the neural network output, and the postprocessing filter output.

4.1   Teeth Detection Experimental Results

A set of 100 images was used. The training and cross-validation set consisted of 10 images, and the test set of 90. Each image had a resolution of 200x200pixels, which means that the training set consisted of 400,000 pixels, and the test set of 3,600,000 pixels.

We wanted to know which image format was best suited for our segmentation task. Therefore, we trained the SVMs and ANNs using each of those formats. RGB, HSV and GRAY refer respectively to the RGB, HSV and gray-scale formats. eRGB and eHSV are the analogous format but with the Xiao&Ohya's enhancement filter, while eRGB', and eHSV' refer to our proposed filter. eGRAY

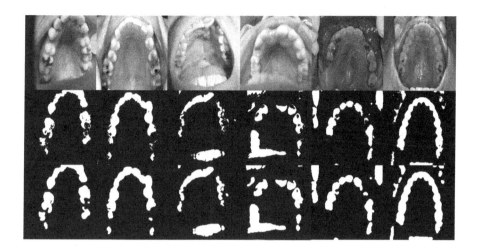

Fig. 3. Teeth Detection. Top: Original Images. Middle: Neural Network output. Bottom: Small blobs filtering.

Table 1. Pixel classification with SVM and ANN

		SVM			ANN			
Arguments	Kernel	Train	Test	Time	Top.	Train	Test	Time
RGB	RBF	45.34	40.40	0.63	50	82.59	80.02	0.10
eRGB'	Linear	52.07	49.86	0.80	100	**86.44**	80.71	0.13
eRGB	Linear	56.59	49.69	0.83	5	83.07	78.47	0.08
HSV	Poly	76.55	72.00	0.63	20	82.78	78.06	0.08
eHSV'	Poly	73.00	55.81	0.27	30	85.85	80.67	0.10
eHSV	RBF	73.69	71.66	0.67	10	85.60	81.44	0.08
GRAY	Linear	63.02	72.32	0.57	50	80.88	**81.79**	0.09
eGRAY	Poly	**82.19**	**79.02**	0.90	5	82.51	80.76	0.06

is the gray-scale format with histogram equalization. Each machine receives as input the format of a pixel.

Table 1 shows that Neural Networks work better in this problem than Support Vector Machines. For each color input format, the best Neural Network found outperformed the best Support Vector Machine found. Moreover, the classification time is much smaller for Neural Networks. An interesting result is that the best SVM found receives as argument the enhanced gray-scale image. Unfortunately, this is the most time consuming SVM too.

At the ANN side, it was hard to tell which was the best network found. The best training classification was achieved using the eRGB' input format. However, this network had 100 hidden neurons. Although it was the most time consuming ANN, notice that it is about 7 times faster than the best SVM found. On the other hand, the GRAY format produced the best classification of test images, yet the worst classification over training.

Table 2. Pixel classification on combined Input Formats for SVM and ANN

		SVM			ANN			
Arguments	Kernel	Train	Test	Time	Top.	Train	Test	Time
eRGB'+eHSV'	Poly	80.89	77.53	0.23	20	**89.02**	81.33	0.10
eRGB + eHSV	Linear	71.02	72.34	0.74	5	88.28	82.59	0.09
eRGB'+eGRAY	Linear	61.57	68.87	0.60	5	86.92	79.44	0.09
eRGB + eGRAY	Poly	80.88	81.79	0.94	30	87.68	80.85	0.10
eHSV'+eGRAY	Poly	75.05	63.30	0.27	25	86.14	80.61	0.09
eHSV + eGRAY	Linear	71.92	71.11	0.69	50	85.41	81.46	0.10
eRGB'+eHSV'+eGRAY	RBF	65.96	44.93	0.83	25	87.84	84.81	0.11
eRGB+eHSV+eGRAY	Linear	80.88	**81.79**	0.77	40	87.55	**85.05**	0.11

Table 3. Teeth Classification Rate After Small Blobs Filtering

Stage	Rate (%)	False Positive (%)	False Negative (%)
eRGB'+eHSV' ANN	81.3300	9.9974	8.6726
Blob Filter	86.1185	5.4553	8.4263

To explore further the impact of the image formats, we combined them, as shown in table 2. In general, significant improvements were achieved for SVMs both in classification rate and time as compared with those of table 1. However, none of these SVMs surpassed the 82.19% of classification rate over training achieved by using the eGRAY format only.

The combination of input image formats improved the classification rates of the Neural Networks as well. The best classification rate obtained over training was 89.02%. This result was achieved by a network of 20 hidden neurons, that combined the RGB and HSV images enhanced with our proposed filter. On the other hand, the best classification rate over testing was achieved by a network that combined the RGB, HSV and GRAY images enhanced by the Xiao&Ohya's filter.

From the results we cannot conclude which is the best Neural Network configuration. However, ANNs outperformed SVMs both in classification rate and time.

Table 3 shows the effect of filtering small blobs from the eRGB'+eHSV' Neural Network output. Blob filter leads to an increment of over 4%, decreasing considerably the false positive rate.

5   Palate Segmentation

Once the teeth regions are found it is easier to determine the palate location. We adjust a curve to the teeth regions by means of optimization algorithms [8]. These algorithms perform rotations, translations, scales and deformations of an original curve learned from palate curves.

When skin pixels are classified, some of them have information indistinguishable from teeth, therefore the neural network classifies them as positive examples. In these cases, the curve adjusting algorithm tries to reach a stable point between skin pixels classified as positive and teeth pixels. Cleaning iteratively the noise located outside the curve, corresponding to skin pixels, leads to better results in this stage. This process ends when fixed point is reached by the skin cleaner method.

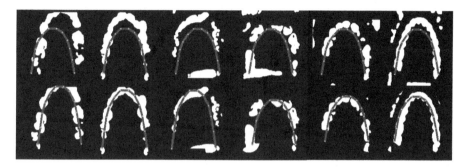

Fig. 4. Curve adjusted and pixel segmentation after filtering pixels out of the curve. Top: Initial curve. Middle: Curve adjusted to the palatal arch. Bottom: Noise filtered.

As can be seen in figure 4, curves are adjusted within teeth pixels. Picture scale, differences between each individual and noise in the classified image may result in different sizes of teeth. It is needed to detect automatically the bandwidth where teeth are included.

Bandwidth detection should decide for each image the right limits where teeth are covered and noise is outside. An unsupervised and supervised combined method is proposed to solve this problem based in the algorithm exposed in [7].

At the unsupervised step, each boundary pixel classified as positive is assigned to a cluster. Assignment of the point p to a cluster c success if the distance between p and c, $d(p, c)$ is less than certain threshold W and $d(p, c) \leq d(p, c')$ for all c' in the set of clusters, where d is an arbitrary function. In our case d is the squared distance between p and the mass center of c. Otherwise, a new cluster is created with the point p.

When every point is assigned, the algorithm classifies each cluster as normal or anomalous data based on the cluster size. Small clusters contain anomalous data, big clusters contain normal data. The bandwidth selected is the biggest mass center of the normal clusters.

The method described is essentially unsupervised, but there are two variables that should be set, W and normal/anomalous cluster threshold. A supervised method that tries to minimize the distance from actual bandwidth to the selected by the unsupervised algorithm is developed as an initial training method.

When the bandwidth is set, points classified positively between the curve and the allowed bandwidth are denoted as teeth. Points inside the polygon generated by the closed curve are denoted as part of palate, as shown in figure 5.

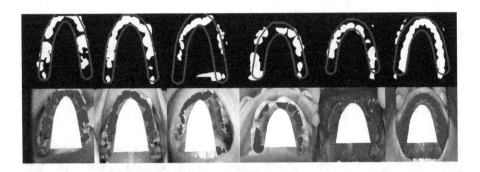

Fig. 5. Bandwidth selection and teeth/palate segmentation

5.1   Bandwidth Selection Experimental Results

Figure 6 shows the mean relative error of the application of bandwidth selection varying W. For each value of W, normal/anomalous threshold is selected such that minimizes the mean error.

The best results found in the training set are values of 13.3 and 0.1 for W and anomalous threshold. With this selection, the mean relative error between the bandwidth selected and the actual bandwidth is 0.09 in the training set and 0.16 in the test set.

Table 4 shows the teeth classification rate variation after the curve adjustment method and bandwidth selection. Recall from table 3 that the Neural Networks classification rate was 81.33%. Small blob filtering combined with the curve adjustment and bandwidth selection methods, improved the teeth classification rate to 89.74%. There is an important decrease in the percentage of false positives, yet there is an increment in the percentage of false negatives because some teeth regions may not be covered by the palatal curve, as it can be seen in figure 5.

Finally, the classification rate of the pixels as palate is 87.5267%. Errors were mainly false positive (8.7598%). We will work on improving the overall performance of our system.

Table 4. Teeth Classification after Curve Adjustment and Bandwidth Selection

Stage	Rate (%)	False Positive (%)	False Negative (%)
Curve Adjustment	87.8218	3.3403	8.8378
Bandwidth Selection	88.0421	1.8916	10.0663

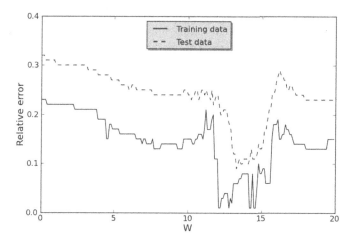

Fig. 6. Bandwidth Selection. Relative error vs. W.

6 Interdental Segmentation

Our first step to segment teeth is to flatten the palatal arch curve, *i.e.*, to transform the teeth band into a rectangle. Unfortunately, some pixels of the resulting rectangle are left blank because of the distortion of the image. Hence, such pixels are filled with a best first algorithm, using as priority the amount of neighboring filled pixels. Each blank pixel is filled with the average color value of its neighbors. Once the rectangle is entirely filled, the image is transformed to gray-scale, and improved by applying histogram equalization [9] [10] and edge enhancement filters [9].

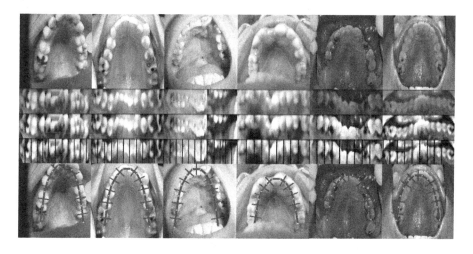

Fig. 7. Interdental Segmentation

Next, a sliding window algorithm searches for boundary between teeth. The main advantage of having the teeth region represented as a rectangle is that the sliding window can have fixed size. The algorithm complexity is linear in the number of window slides times the complexity time of the classifier.

We tested Support Vector Machine [12] and Multilayer Perceptron trained with Backpropagation [11]. Input are image slices of 50×25 pixels. The classifiers are trained to detect gaps between teeth.

6.1   Interdental Segmentation Experimental Results

ANN outperform SVM at this task as shown in table 5. Best results in time and quality were reached with Neural Networks. The best Neural Network found had 50 neurons at the hidden layer and worked with the original flattened image. A smaller Neural Network of 15 hidden neurons reached the same classification rate during testing using the enhanced images as input. However, this networks did not reach perfect classification rate during training.

Table 5. Interdental Segmentation

		SVM				ANN		
Arguments	Kernel	Train	Test	Time (ms)	Top.	Train	Test	Time (ms)
Original	Linear	86.82	81.14	149	50	**100.00**	**88.82**	85
Enhanced	Poly	**92.27**	**83.99**	154	15	99.55	88.82	92

7   Conclusions and Future Work

We presented our results on automatic teeth/palate and interdental segmentation in palate view photographs of the upper jaw. Classic Multilayer Perceptrons trained with the error backpropagation algorithm outperformed SVMs for both of these tasks.

Image enhancement, filtering and adaptive curve fitting helped differentiate teeth and palates. Our system was able to cope with variations in patient anatomy, mouth scale in pictures, and view angle. Likewise, it was able to overcome standard problems for image processing systems, such as differences in illumination and image quality.

We wish to make our system more robust, so we are extending its capabilities to detect whether or not the input image depicts the upper jaw. Our final goal is to effectively detect common oral infecto-contagious diseases by looking at the palate, which could help diagnose HIV and AIDS. We believe this is an important initial step in that direction.

Acknowledgment. We would like to thank Dr. Vilma Tovar from the "Centro de Atención a Personas con Enfermedades Infectocontagiosas Dra Elsa La Corte", Universidad Central de Venezuela. Dr. Tovar provided us with photographies of oral diseases and their diagnose.

References

1. World Health Organization: Progress report 2011: Global HIV/AIDS response Epidemic update and health sector progress towards universal access. WHO Press, ISBN 978 92 4 150298 6,
   `http://www.who.int/hiv/pub/progress_report2011/en/index.html`
2. Barr, C.E.: Dental management of HIV-associated oral mucosal lesions: current and experimental techniques. In: Robertsonn, P.B., Greenspan, J.S. (eds.) Perspectives on Oral Manifestation of AIDS: Diagnosis and Management of HIV-Associated Infections, pp. 77–95. PSG Publishing Co., Inc., Littleton (1988)
3. Arendorf, T.M., Bredekamp, B., Cloete, C., Sauer, G.: Oral manifestation of HIV infection in 600 South African patients. J. Oral Pathol. Med. 27, 176–179 (1998)
4. Nassar, D., Ammar, H.: A neural network system for matching dental radiographs. Pattern Recognition 40, 65–79 (2006)
5. Lin, P., Lai, Y., Huang, P.: An effective classification and numbering system for dental bitewing radiographs using teeth region and contour information. Pattern Recognition (2009)
6. Xiao, D., Ohya, J.: Contrast Enhancement of Color Images Based on Wavelet Transform and Human Visual System. In: Proceedings of the IASTED International Conference Graphics and Visualization in Engineering, Florida (2007)
7. Chimphlee, C., Hanan, A., Noor, M.: Unsupervised Anomaly Detection with Unlabeled Data Using Clustering. In: Proceedings of the Postgraduate Annual Research Seminar (2005)
8. Gendreau, M., Potvin, J.: Handbook of Metaheuristics. Springer (2010)
9. Gonzalez, R., Woods, R.: Digital Image Processing, 3rd edn. Prentice Hall (2007)
10. Shapiro, L., Stockman, G.: Computer Vision. Prentice Hall (2001)
11. LeCun, Y., Bottou, L., Orr, G.B., Müller, K.-R.: Efficient BackProp. In: Orr, G.B., Müller, K.-R. (eds.) Neural Networks: Tricks of the Trade. LNCS, vol. 1524, pp. 9–50. Springer, Heidelberg (1998)
12. Burges, C.: A tutorial on support vector machines for pattern recognition. Knowledge Discovery and Data Mining 2(2) (1998)
13. Li, M., Cheung, Y.-M.: Automatic Segmentation of Color Lip Images Based on Morphological Filter. In: Diamantaras, K., Duch, W., Iliadis, L.S. (eds.) ICANN 2010, Part I. LNCS, vol. 6352, pp. 384–387. Springer, Heidelberg (2010)

On Instance Selection
in Audio Based Emotion Recognition

Sascha Meudt and Friedhelm Schwenker

University of Ulm
Institute of Neural Information Processing
89069 Ulm, Germany
{Sascha.Meudt,Friedhelm.Schwenker}@uni-ulm.de
http://www.uni-ulm.de/in/neuroinformatik.html

Abstract. Affective computing aim to provide simpler and more natural interfaces for human-computer interaction applications, e.g. recognizing automatically the emotional status of the user based on facial expressions or speech is important to model user as complete as possible in order to develop human-computer interfaces that are able to respond to the user's action or behavior in an appropriate manner. In this paper we focus on audio-based emotion recognition. Data sets employed for the statistical evaluation have been collected through Wizard-of-Oz experiments. The emotional labels have been are defined through the experimental set up therefore given on a relatively coarse temporal scale (a few minutes) which This global labeling concept might lead to miss-labeled data at smaller time scales, for instance for window sizes uses in audio analysis (less than a second). Manual labeling at these time scales is very difficult not to say impossible, and therefore our approach is to use the globally defined labels in combination with instance/sample selection methods. In such an instance selection approach the task is to select the most relevant and discriminative data of the training set by using a pre-trained classifier. Mel-Frequency Cepstral Coefficients (MFCC) features are used to extract relevant features, and probabilistic support vector machines (SVM) are applied as base classifiers in our numerical evaluation. Confidence values to the samples of the training set are assigned through the outputs of the probabilistic SVM.

Keywords: Emotion Recognition, Human Computer Interaction, Instance Selection, Active Learning.

1 Introduction and Motivation

In supervised learning a large amount of labeled training data has to be collected in order to construct models of acceptable prediction accuracy, and so in pattern recognition or data mining application the training set design is one of the most important parts of the overall process. Designing a training set means pre-processing the raw data, selecting the relevant features, selecting the representative instances (samples), and labeling the samples for application at hand.

N. Mana, F. Schwenker, and E. Trentin (Eds.): ANNPR 2012, LNAI 7477, pp. 186–192, 2012.
© Springer-Verlag Berlin Heidelberg 2012

Labeling data is time consuming, expensive (e.g. at least in cases where more than one expert must be asked), and of course error-prone. On the other hand, in many pattern recognition applications, such as classification of text documents, remote sensing, or image/video classification, big pools of unlabeled data are available [5,17].

Emotion classification from audio data is a challenging pattern recognition task. Experiments on acted emotional data sets show the human perception capability is similar to the performance of automatic classifiers, particularly humans produce high error rates on emotional data sets, higher than in many other recognition tasks [4]. Labeling emotional data sets is extremely difficult and time consuming, and therefore specific annotation tools must be applied in this task [12], in particular when real world emotional data has to be analyzed, i.e. the emotional utterances are naturalistic emotions in real human-computer-interaction (HCI) scenarios, for instance in Wizard-of-Oz settings [6,19]. In naturalistic HCI scenarios emotional utterances are mainly *neutral*, typically only a few low intensity emotional patterns can be observed in such data streams. The annotation of such WoZ data is usually driven through the experimental design.

Instance selection deals with searching for a small subset S of the original training set T, such that a classifier trained on S shows similar, or even better classification performance than a classifier trained on the full data set T [9,2,11,7]. We will present confidence-based instance selection criteria for probabilistic support vector machines based on cross-validation. The statistical evaluation of the proposed selection method has been performed on task of affect recognition from speech. Classes are not defined very well in this type of application and therefore lead to data sets with high label noise. Numerical evaluations on these data sets show that classifiers can benefit from instance selection not only in terms of computational costs but even in terms of classification accuracy [3,15,10,13].

The paper is organized as follows: In section 2 the data set and feature extraction procedure are briefly described, then in section 3 an overview on the base classifiers is given. Results are discussed in section 4 and a preliminary conclusion with future work is given in section 5.

2   Data Collection and Feature Extraction

The data used to validate the architecture was collected in a Wizard-of-Oz experiment where human-computer interaction (HCI) is simulated [8]. Within the study, the computer interacts as a mental trainer of the popular game "Concentration" while the subjects are able to control the system using short speech commands. The setup induces emotions according to the Valence-Arousal-Dominance (VAD) model [16] using the following affective factors:

- Delaying the response of a command
- Non-execution of the command
- Simulating incorrect speech recognition

- Offering technical assistance
- Lack of technical assistance
- Propose to quit the game ahead of time
- Positive feedback

The procedure of emotion induction is structured in different experimental sequences (ES-4 and ES-6) in which the user is passed through VAD octants by the investigator. Within this study we focus on the recognition of the emotional octants in ES-4 and ES-6 (positive valence, low arousal, high dominance versus negative valence, high arousal, low dominance). The database used consists of 6 subjects with an average age of 63.5 years. Audio, video and physiological data was recorded. In this approach only the audio part was used for classification.

In order to extract the speech non-speech segments from the recorded audio, an energy-based threshold was defined. The energy was determined using a window of 40ms. From this signal, Mel-Frequency Cepstral Coefficients (MFCC) features with a 20 dimensional filter bank were calculated [18]. The windows are shifted with an offset of half the respective window size. This results in about 3000 feature vectors per individual, which in average are equally balanced on the both experimental parts.

3   Methodology

In our architecture we propose the utilization of a Support Vector Machine (SVM). The SVM is a supervised learning method following the maximum margin paradigm. The kernel trick increases the dimensionality of the feature space and therefore allows non-linear non-linear separation surfaces. Within our study we used the Gaussian Radial Basis Function (RBF) kernel, which transforms the input data into the Hilbert space of infinite dimensions and is calibrated by the parameter γ. Due to noise or wrong annotations it is convenient to have a non-rigid hyper-plane, being less sensitive to outliers in training. Therefore, an extension to the SVM introduces a so-called slack term that tolerates the amount of misclassified data using the control parameter C. A probabilistic classification output can be obtained using the method proposed by Platt et al. [14]. Detailed information of the algorithm can be found for instance in [1].

Instance selection Algorithm

```
Input: Dataset T, Reduction amount r, Number of classes l

split Dataset T into N bags
FOR EACH t_i in T (i=1..N)
    train SVM on T without t_i
    classify each x in t_i
END FOR

remove all misclassified examples x from T
```

```
build S by taking most confident r/l examples x of each class

train SVM on S

Output: Reduced Dataset S, SVM
```

First we use a cross-validation based approach to reclassify the complete dataset. We split the original feature dataset T into N bags, using $N - 1$ bags for training a probabilistic SVM. This SVM then is used to reclassify the left out bag and mark each feature vector in it with its reclassification decision and confidence. These procedure is repeated for each of the N bags. After marking all feature vectors of the original dataset, the dataset is reduced by keeping only the highest confident and correct reclassified instances. In addition a balancing constraint of the two classes is achieve by taking the same amount of features of each class, even if this implies that the chosen confidence thresholds for the two classes differ. The reduced dataset S then is used to train a second stage classifier where we again chose a probabilistic SVM.

Due to the fact that emotion recognition is a highly individual task, the instance selection and SVM training task is done separately for each individual.

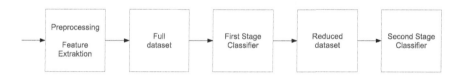

Fig. 1. Architecture of the instance selection based architecture

4   Statistical Evaluation

In this section, the statistical evaluation of the instance selection architecture is described for the above mentioned data. First the first stage classifier is evaluated which is used for the selection. In a second part the second stage classifiers that where trained with the reduced dataset are evaluated. We used three different reduction intensities of 98%, 90% and 80% reduction from the original dataset. The reduced dataset contains an equal balanced amount of feature vectors from both classes. The SVM parameters where set to $C = 2$ and $\gamma = 4 \cdot 10^{-3}$. Each individual result was evaluated by a 5 fold cross-validation procedure. The individual results where combined by averaging the individual results.

In case of the default classification approach without reduction or rejection we reached a classification accuracy of 65.9% and a F1 measure of 0.628. Adding a rejection option of at least 0.95 classification confidence the classification accuracy increases to 81.8% (F1 0.790). The amount of rejection is nearly linear to the confidence threshold. This means that in case of 0.95 minimum confidence about 90% of the decisions had been rejected.

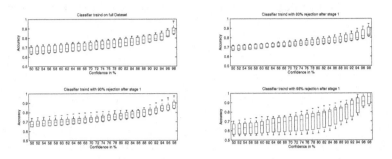

Fig. 2. Accuracy depending on confidence threshold of rejection. Average of all subjects.

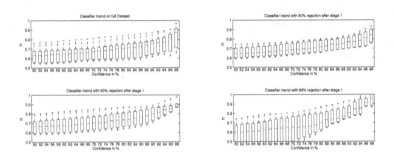

Fig. 3. F1 measure depending on confidence threshold of rejection. Average of all subjects.

Adding the second stage trained with a 80% reduced dataset improved this value. In case of rejectionless classification we reached a classification accuracy of 68.5% with nearly half standard derivation compared to the classifier trained on the whole dataset. Also the F1 measure got up to 0.681 with better standard derivation.

Reducing the dataset to 2% of the whole set and adding a rejection option with minimum confidence of 0.95 we achieved the best result. The classification accuracy got up to 96% also the F1 measure improved to 0.965. In addition the confidence derivation got better, more SVM classification decisions got high confidence values. In case of rejection decision with less than 0.95 confidence we got nearly twice the amount of decisions compared with the not reduced classifier. Keeping in mind that the feature extraction is based on a shifted window with offset of a half window, this implies that around 40% of the time-line is covered with decisions. Or on the other hand the confidence threshold could be set up to nearly 0.9 (Accuracy 87%) by still covering nearly the whole time-line with decisions.

In all cases we could find an instance selection based architecture that outperforms the well known standard SVM training approach in all cases (Accuracy, F1 measure, Standard Derivation and Confidence Derivation). Finally the computational cost for training and classification on a heavy reduced dataset is highly decreased compared to training on a large dataset with lots of noisy data in it.

5   Conclusion and Future Work

Classifying the emotion is generally a difficult task when leaping from overacted data to realistic human computer interaction. In this study the problem was investigated with respect to the fact that only less parts of the dataset contain intense emotion. The result of the evaluation is that the usage of instance selection can reduce the testing error, or reducing the standard derivation depending on the chosen parameters.

Rejecting samples when classifying such kind of data turns out to be a sound approach. Especially when the distribution of the classes in the data is heavily overlapping. For future work, it could be promising to implement an iterative classifier training procedure, were the training data can be rejected.

Using more than just the audio part of the dataset could also improve the results, by using a co training multi classifier approach.

Acknowledgments. This research was supported in part by grants from the Transregional Collaborative Research Centre SFB/TRR 62 "Companion-Technology for Cognitive Technical Systems" funded by the German Research Foundation (DFG).

References

1. Bishop, C.: Pattern recognition and machine learning, vol. 4. Springer, New York (2006)
2. Brighton, H., Mellish, C.: Advances in instance selection for instance-based learning algorithms. Data Mining and Knowledge Discovery 6(2), 153–172 (2002)
3. Domingo, C., Gavaldà, R., Watanabe, O.: Adaptive sampling methods for scaling up knowledge discovery algorithms. Data Mining and Knowledge Discovery 6(2), 131–152 (2002)
4. Esparza, J., Scherer, S., Brechmann, A., Schwenker, F.: Automatic emotion classification vs. human perception: Comparing machine performance to the human benchmark. In: International Conference on Information Science, Signal Processing and Their Applications (ISSPA 2012), pp. 1286–1291 (2012)
5. Esparza, J., Scherer, S., Schwenker, F.: Studying Self- and Active-Training Methods for Multi-feature Set Emotion Recognition. In: Schwenker, F., Trentin, E. (eds.) PSL 2011. LNCS, vol. 7081, pp. 19–31. Springer, Heidelberg (2012)
6. Glodek, M., Tschechne, S., Layher, G., Schels, M., Brosch, T., Scherer, S., Kchele, M., Schmidt, M., Neumann, H., Palm, G., Schwenker, F.: Multiple classifier systems for the classification of audio-visual emotional states. In: 1st International Audio/Visual Emotion Challenge and Workshop (2011)
7. de Haro-García, A., García-Pedrajas, N., del Castillo, J.A.R.: Large scale instance selection by means of federal instance selection. Data Mining and Knowledge Engineering 75, 58–77 (2012)

8. Kelley, J.: An iterative design methodology for user-friendly natural language office information applications. ACM Transactions on Information Systems (TOIS) 2(1), 26–41 (1984)

9. Liu, H., Motoda, H.: Instance Selection and Construction for Data Mining. Kluwer Academic Publishers, Norwell (2001)

10. Liu, H., Motoda, H.: On issues of instance selection. Data Mining and Knowledge Discovery, 115–130 (2002)

11. Madigan, D., Raghavan, N., DuMouchel, W., Nason, M., Posse, C., Ridgeway, G.: Likelihood-based data squashing: A modeling approach to instance construction. Data Mining and Knowledge Discovery 6(2), 173–190 (2002)

12. Meudt, S., Bigalke, L., Schwenker, F.: ATLAS – an annotation tool for HCI data utilizing machine learning methods. In: Proceedings of the 4th International Conference on Applied Human Factors and Ergonomics, AHFE 2012 (in print, 2012)

13. Olvera-Lpez, J.A., Carrasco-Ochoa, J.A., Trinidad, J.F.M., Kittler, J.: A review of instance selection methods. Artificial Intelligence Reviews, 133–143 (2010)

14. Platt, J.: Probabilistic outputs for support vector machines and comparisons to regularized likelihood methods. Advances in Large Margin Classifiers 10(3), 61–74 (1999)

15. Reinartz, T.: A unifying view on instance selection. Data Mining and Knowledge Discovery 6(2), 191–210 (2002)

16. Russell, J.A., Mehrabian, A.: Evidence for a three-factor theory of emotions. Journal of Research in Personality 11(3), 273–294 (1977)

17. Schels, M., Kächele, M., Hrabal, D., Walter, S., Traue, H.C., Schwenker, F.: Classification of Emotional States in a Woz Scenario Exploiting Labeled and Unlabeled Bio-physiological Data. In: Schwenker, F., Trentin, E. (eds.) PSL 2011. LNCS, vol. 7081, pp. 138–147. Springer, Heidelberg (2012)

18. Scherer, S., Glodek, M., Schwenker, F., Campbell, N., Palm, G.: Spotting laughter in natural multiparty conversations: A comparison of automatic online and offline approaches using audiovisual data. TiiS 2(1), 4 (2012)

19. Walter, S., Scherer, S., Schels, M., Glodek, M., Hrabal, D., Schmidt, M., Böck, R., Limbrecht, K., Traue, H.C., Schwenker, F.: Multimodal Emotion Classification in Naturalistic User Behavior. In: Jacko, J.A. (ed.) HCII 2011, Part III. LNCS, vol. 6763, pp. 603–611. Springer, Heidelberg (2011), http://www.springerlink.com/content/606237v0u5225w50/

Traffic Sign Classifier Adaption
by Semi-supervised Co-training

Matthias Hillebrand[1], Ulrich Kreßel[1], Christian Wöhler[2], and Franz Kummert[3]

[1] Daimler AG, Group Research and Advanced Engineering, 89081 Ulm, Germany
{matthias.hillebrand,ulrich.kressel}@daimler.com
[2] Image Analysis Group, TU Dortmund, 44221 Dortmund, Germany
christian.woehler@tu-dortmund.de
[3] Applied Informatics Group, Bielefeld University, 33615 Bielefeld, Germany
franz@techfak.uni-bielefeld.de

Abstract. The recognition of traffic signs in many state-of-the-art driver assistance systems is performed by statistical pattern classification methods. Traffic signs in European countries share many similarities but also vary in colour, size, and depicted symbols, making it hard to obtain one general classifier with good performance in all countries. Training separate classifiers for each country requires huge amounts of labelled training data. A well-trained classifier for one country can be adapted to other countries by semi-supervised learning methods to perform reasonably well with relatively low requirements regarding labelled training data. Self-training classifiers adapting themselves to unknown domains always risk that the adaption will become ineffective or even fail completely due to the occurrence of incorrectly labelled samples. To assure that self-training classifiers adapt themself correctly, advanced multi-classifier training methods like co-training are applied.

Keywords: self-training, semi-supervised, co-training.

1 Introduction

The automatic recognition of traffic signs by a driver assistance system, a mature but still contemporary field of research, commonly utilises intensity-based classifiers for processing visual sensor information. In classical supervised learning-based approaches, a classifier has to be trained for each country separately, even if the traffic signs in European countries reveal only minor variations regarding colour, font, font size and depicted symbols.

Such a supervised approach is inefficient because it leads to high labelling costs for human annotators, while unlabelled data can often be acquired with justifiable expenditure. Since there are not too many variations between the different countries, an existing classifier for one country can be extended using informative data samples from other countries. An automatic class assignment is desirable to further reduce the labelling costs. In literature, this approach is known as semi-supervised learning.

N. Mana, F. Schwenker, and E. Trentin (Eds.): ANNPR 2012, LNAI 7477, pp. 193–200, 2012.

Individual semi-supervised classifiers rely on their own capabilities to predict labels correctly and learn these predictions to extend their knowledge or rather their classification competence. But this approach risks the adapted classifier not improving in performance or even becoming ineffective due to the self-teaching of incorrect labelled samples. Combining two or more classifiers with different architectures may help to represent a more comprehensive variability of the domain to be learned and stabilise of the learning process.

This paper proposes an iterative co-training process where the most informative samples from a given pool of unlabelled traffic sign images are automatically selected and then classified by two classifiers which predict labels for each other. Additional knowledge of traffic sign appearance in countries unknown to the classifiers as well as noise distributions in rotation, camera angle, etc., will be obtained from a set of synthetically generated samples.

The evaluation starts with self-teaching a classifier for Germany, beginning with a relatively small initial training set. The adaption of this self-taught German classifier on Italian traffic signs without any intervention by a human expert is evaluated further as well as an adaption of a fully supervised trained German classifier. The classifiers distinguish between 14 different classes of traffic signs including speed limits, no-passing signs, and the corresponding termination signs. In the following sections, the proposed method is elaborated in detail.

2   Related Work and Applied Methods

The traffic sign recognition survey by Fu and Huang [5] provides an introduction and a brief overview of the broad field of existing approaches. A recent and comprehensive publication on automatic traffic sign recognition and classification by Lindner [10] describes a complete processing chain including the recording of training data, the detection and tracking of traffic signs, the training and evaluation of classifiers and finally the application of heuristics to bring the classification results into accordance with other traffic rules in order to construct a speed limit driver assistance system. Today, traffic sign recognition systems are available as special equipment for some car models from renowned manufacturers, but none of these systems has self-learning capabilities.

Semi-supervised learning techniques [14],[3],[13] reduce the amount of required labelled samples when training new traffic sign classifiers or adapt existing classifiers to other countries as described by Hillebrand et. al [7]. Combining two or more classifiers to obtain better classification results or to further reduce the data acquisition and labelling efforts with the help of active learning methods where applicable can be achieved by Blum and Mitchell's co-training approach [1] or by means of ensemble learning methods [11]. Recently, Cui et al. [4] analysed the application of co-training methods in semi-supervised two-class traffic sign learning scenarios.

Like Cui et al., we co-train two different classifiers, a polynomial classifier (PC) [12] with a fully quadratic structure of the feature vectors and a support vector regression (SVR) [2]. The polynomial classifier is robust against partially mislabelled data and therefore capable of producing good classification results even if the semi-supervised learning process associates a certain fraction of samples with incorrect class labels. In contrast to this, the support vector regression is more sensitive towards mislabelled samples and just produces an exceeding amount of support vectors without achieving generalisation capability for the classification task, but is better in representing special cases not considered by a PC model function. The n-class classification capabilities are realised by n one-against-all epsilon-SVRs with radial basis functions.

Co-training relies on the assumption that two different classifiers can represent a more comprehensive variability of the feature space due to their different model hypotheses. The semi-supervised training is an iterative process. After initialising the classifiers with a certain amount of correctly labelled samples both classifiers predict labels for a certain amount of selected samples from the large set of available unlabelled samples during each iteration. It is preferable to select those samples to be classified which are the most informative ones or rather let the classifier make a lunge towards ground truth performance. On the other hand, only samples the classifier is capable of predicting a confident label for can be selected. The confidence bands method [6],[7] is used to obtain a balanced selection of samples. Confidence bands are curves enclosing a model function estimated by a regression analysis. The bands represent the areas where the true model is expected to reside with a certain probability, commonly 95%. The extent of the bands in different areas of the data space gives an idea of how well the estimated function fits the data. Each iteration finishes by adding the samples classified by classifier A along with their predicted labels to the training set of classifier B and vice versa.

Real images of traffic signs depict a wide spectrum of variations, e.g., different sizes, rotations and camera angles, translations due to inaccurate detection results, soiling, partial occlusions, or different lighting and weather conditions. Lighting conditions are normalised by a pre-processing algorithm. The other most common variations (size, rotation, camera angle, translation) are represented by virtual traffic sign samples which are generated from one ideal depiction of each traffic sign by the method described by Hoessler et al. [8], but the less frequent variations are unrepresented. In principle, an infinite amount of such virtual traffic sign samples are available. The virtual samples are utilised to support the semi-supervised learning process and give the classifiers a basic training for unknown domains.

3 Experimental Evaluation

3.1 Experimental Setup

Applying the pre-processing steps described in detail in [7] results in intensity images with a size of 17×17 pixels and adjusted lighting conditions displaying traffic sign and garbage cut-outs. The automatic traffic sign vs. garbage sample classification leads to a rate of about 5% garbage samples in the training set. This complicates a lasting semi-supervised classifier training.

Fig. 1. Visualised PCA and NMF basis vectors with size of 17 × 17 pixels. Left: PCA basis vectors, known as *eigenfaces* in face recognition applications; black pixels are negative values, white pixels are positive values. Right: NMF basis vectors; grey pixels are zero values, white pixels are positive values.

The intensity values are used as a feature vector of dimension 289 reduced to 23 by a Principal Component Analysis preserving about 80% of the image information. In general, co-training assumes that each sample is represented using two different feature sets that provide complementary information. To get a second view on the data for the second classifier, a non-negative matrix factorisation (NMF) [9] with the same feature dimension and a comparable mean reconstruction error on the data samples is computed. The most significant difference between the two matrix factorisation methods is that the NMF results in basis images and sample encodings with only elements equal to or greater than zero (see Fig. 1). In all experiments, the PC classifier will be trained with the NMF-reduced feature vectors and the SVR will be trained with the PCA-reduced feature vectors.

The evaluation applies the self-teaching classifiers to different learning scenarios. The performance evaluation starts with self-teaching three classifiers for German traffic signs, a PC classifier, a SVR classifier and a co-training classifier consisting of both individual classifiers. The learning process starts with a relatively small initial training set which contains 10% of the whole set. These samples are taken randomly from the whole set, but are distributed equally over all classes. The whole German data set consists of 14 classes (see Fig. 2) containing 1000 samples per class (14000 samples in total). A training set of the same composition and the same size is available from Italy. Finally, one set of virtual training samples, again with 14 classes and 14000 samples in total, is generated for each country.

The second part of the evaluation compares the performance of the three classifiers described above to a German-Italian adaption scenario: The self-trained co-training classifier from the first evaluation as well as a classifier trained with the same training set but with ground truth labels will be adapted to recognise Italian traffic signs by self-teaching samples from the Italian training set iteratively.

All evaluations consisting of random initial training sets have been repeated five times. The resulting classification rates given in the tables are mean values. The standard deviation is always included as well, except for classifiers with initial training sets consisting of all virtual samples.

The classifier performances are compared by computing correct classification rates on independent test sets. Like the training sets, the test sets consist of samples from the same 14 classes with each class represented by 250 samples (3500 samples in total). The test sets have been recorded independently of the

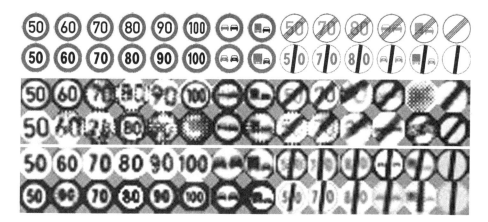

Fig. 2. Traffic sign images. Rows 1–2: Ideal depictions of 14 German and Italian signs. Rows 3–4: Two German real training samples from each class. Rows 5–6: Two Italian virtual training samples from each class.

training sets with different cameras. The same pre-processing routines have been applied to all samples from all data sets.

The data sets are not noise-free and contain a considerable amount (between 5% and 10%) of garbage samples and bad quality samples, e.g., with inaccurate cut-outs. An example of this is depicted in Fig. 2 for some German training samples. Classifiers often obtain correct classification rates above 90%, so the large number of bad quality samples in the test sets would not allow meaningful comparisons between different classifiers. For this reason, all samples not classifiable by a human expert were removed from the test sets before the 3500 samples per country were selected at random.

3.2   Reference Classifiers

The evaluation begins with the determination of reference classifiers to benchmark the performances of all following self-trained classifiers. In general, the Italian reference classifiers perform about five percentage points below the German classifiers as shown in Table 1. All classifiers trained with real samples always show the best performance. The classifiers trained with real and virtual samples perform on the same level or a little bit below. The classifiers trained only with virtual samples perform several percentage points worse. This result shows that there is no need to use virtual training samples when a sufficient amount of labelled real training samples are available.

The co-training classifiers, which combine polynomial classifier and support vector regression, perform comparable to the SVR classifiers alone. Only the PC classifiers perform somewhat worse than the other. Further investigations show that the comparatively poor performance is at least in part a consequence of the 5–10% amount of garbage samples in the training sets.

Table 1. Correct classification rates of all reference classifiers, determined on country-specific test sets. Three different German (rows 1–3) resp. Italian (rows 4–6) training sets contain only real samples, real and virtual samples or only virtual samples.

classifier	real samples	real & virtual samples	virtual samples
DE: PC	0.98	0.96	0.86
DE: SVR	0.99	0.99	0.90
DE: PC & SVR	0.99	0.98	0.89
IT: PC	0.93	0.91	0.81
IT: SVR	0.94	0.94	0.83
IT: PC & SVR	0.94	0.93	0.83

3.3 Self-teaching German Traffic Signs

This evaluation is about self-teaching a German classifier with a very small amount of labelled real training samples or even no labelled real samples at all. All three classifiers presented in Table 2, PC, SVR and PC & SVR cotraining, are trained with 10% real training samples followed by self-teaching the other 90%, with 10% real and all virtual training samples followed by self-teaching the other 90% real samples, and finally with only virtual initial training samples, followed by self-teaching all 100% real samples. The classification rates are determined on the German test set.

As expected, all classifiers show their best performances for the second scenario, followed by the third scenario. The classifier from the first scenario with only 10% real initial samples (these 10% are the only samples that are definitely correctly labelled) are still close to the other classifiers' performances (except the SVR). When inspecting the standard deviations of the classification rates, the co-training classifiers which show fewer correct classifications than the PC classifiers also result in a slightly lower standard deviation.

Table 2. Correct classification rates of German self-teaching classifiers. The initial training sets contain 10% real samples, all virtual and 10% real samples or only virtual samples. The classifiers always self-taught the rest of the real German samples.

initial training set classifier	10% real		virtual & 10% real		only virtual
	class. rate	std. dev.	class. rate	std. dev.	class. rate
PC	0.94	0.017	0.95	0.015	0.94
SVR	0.81	0.041	0.91	0.033	0.91
PC & SVR	0.93	0.011	0.95	0.012	0.93

3.4 Adapting German Classifiers to Italy

This evaluation is about adapting existing classifiers trained on the complete set of German real traffic signs to Italy. The classification rates are determined on the Italian test set. As a first experiment, all three classifiers (PC, SVR and PC & SVR co-training) obtain 10% real training samples from Italy with correct labels to their training sets, followed by self-teaching the other 90%. In the second experiment, 10% real and all virtual training samples from Italy are mixed into their training sets, followed by self-teaching the other 90% real samples. Finally the classifiers are equipped only with the complete set of Italian virtual training samples, followed by self-teaching all 100% real samples.

As Table 3 shows, the best classification rates are obtained by the PC classifiers, again. But the co-training classifiers also perform reasonably well with a slightly lower deviation in their results, again. As expected after the previous evaluations, the SVRs alone perform worst of all. When using the self-trained German classifier which incorporates about 9% incorrectly labelled samples as a basic training, the performance decreases by some percentage points.

Table 3. Correct classification rates of German classifiers adapted to Italy. Rows 1–3: The classifiers have been trained initially with all German real samples. Row 4: The classifier has been semi-supervised co-trained with German training samples (contains incorrectly classified samples). Initial training sets also contain Italian samples: 10% real samples, all virtual and 10% real samples or only all virtual samples. The classifiers always self-taught the rest of the real Italian samples.

initial training set	10% real		virtual & 10% real		only virtual
classifier	class. rate	std. dev.	class. rate	std. dev.	class. rate
PC	0.89	0.032	0.90	0.028	0.81
SVR	0.83	0.038	0.85	0.040	0.79
PC & SVR	0.88	0.022	0.88	0.019	0.81
PC & SVR (DE-st)	0.81	0.056	0.82	0.051	0.77

4 Summary and Conclusions

Adapting a German traffic sign classifier to Italy by semi-supervised methods shows a reasonable performance. Although not discussed in this work, an adaptation to other European countries like Austria, Spain or France yields comparable results.

The main conclusion of this work is that classifiers with the best performances should not always be preferred for self-training tasks. In the majority of cases, the co-trained classifier does not show the best mean classification rates but the lowest standard deviation of the performance measure. This means that using this kind of classifier for a self-training task, the resulting classifier may not be the best, but supplies a reasonable performance with a considerably higher confidence.

Adding virtual training samples to the initial training sets always increases traffic sign classifier performances or stabilises the results. Due to the theoretically infinite availability of these samples, it appears reasonable to always include them in the training sets.

All classifier training runs were processed until 100% of the training samples were labelled by the classifiers and added to the training sets. A result not revealed by the tables in the evaluation section is that many classifiers show their best performance after self-teaching a smaller amount of samples, sometimes even less than 50%. The performance of some classifiers eventually decreased slightly due to an increasing amount of incorrectly labelled samples. An early stopping criterion is therefore being developed.

References

1. Blum, A., Mitchell, T.: Combining labeled and unlabeled data with co-training. In: Proceedings of the Eleventh Annual Conference on Computational Learning Theory, COLT 1998, pp. 92–100. ACM, New York (1998)
2. Chang, C.C., Lin, C.J.: LIBSVM: A library for support vector machines. ACM Transactions on Intelligent Systems and Technology 2, 27:1–27:27 (2011)
3. Chapelle, O., Schölkopf, B., Zien, A. (eds.): Semi-Supervised Learning. Adaptive Computation and Machine Learning. The MIT Press (2006)
4. Cui, T., Grumpe, A., Hillebrand, M., Kreßel, U., Kummert, F., Wöhler, C.: Analytically tractable sample-specific confidence measures for semi-supervised learning. In: Proc. 21st Workshop Computational Intelligence, pp. 171–186 (2011)
5. Fu, M.Y., Huang, Y.S.: A survey of traffic sign recognition. In: 2010 International Conference on Wavelet Analysis and Pattern Recognition (ICWAPR), pp. 119–124 (2010)
6. Hillebrand, M., Wöhler, C., Krüger, L., Kreßel, U., Kummert, F.: Self-learning with confidence bands. In: Proc. 20th Workshop Computational Intelligence, pp. 302–313 (2010)
7. Hillebrand, M., Wöhler, C., Kreßel, U., Kummert, F.: Semi-supervised Training Set Adaption to Unknown Countries for Traffic Sign Classifiers. In: Schwenker, F., Trentin, E. (eds.) PSL 2011. LNCS, vol. 7081, pp. 120–127. Springer, Heidelberg (2012)
8. Hoessler, H., Wöhler, C., Lindner, F., Kreßel, U.: Classifier training based on synthetically generated samples. In: The 5th International Conference on Computer Vision Systems (2007)
9. Lee, D.D., Seung, H.S.: Learning the parts of objects by non-negative matrix factorization. Letters to Nature 401(1), 788–791 (1999)
10. Lindner, F.: Adaptive Traffic Sign Recognition. Ph.D. thesis, Bielefeld University (2012)
11. Rokach, L.: Pattern Classification using Ensemble Methods. Series in Machine Perception and Artificial Intelligence, vol. 75. World Scientific (2010)
12. Schürmann, J.: Pattern Classification: A Unified View of Statistical and Neural Approaches. John Wiley & Sons, Inc. (1996)
13. Xu, Z., King, I., Lyu, M.R.: More Than Semi-supervised Learning. Lambert Academic Publishing (2010)
14. Zhu, X., Goldberg, A.B.: Introduction to Semi-Supervised Learning. Synthesis Lectures on Artificial Intelligence and Machine Learning. Morgan & Claypool Publishers (2009)

Using Self Organizing Maps to Analyze Demographics and Swing State Voting in the 2008 U.S. Presidential Election

Paul T. Pearson[1] and Cameron I. Cooper[2]

[1] Hope College, PO Box 9000, Holland, MI 49422, USA
paultpearson@gmail.com
[2] Fort Lewis College, 1000 Rim Drive, Durango, CO 81301, USA
Cooper_C@fortlewis.edu
http://faculty.fortlewis.edu/Cooper_C

Abstract. Emergent self-organizing maps (ESOMs) and k-means clustering are used to cluster counties in each of the states of Florida, Pennsylvania, and Ohio by demographic data from the 2010 United States census. The counties in these clusters are then analyzed for how they voted in the 2008 U.S. Presidential election, and political strategies are discussed that target demographically similar geographical regions based on ESOM results. The ESOM and k-means clusterings are compared and found to be dissimilar by the variation of information distance function.

Keywords: Kohonen self organizing map, k-means clustering, variation of information, United States election 2008, United States Census data 2010.

1 Introduction

The United States presidential election in 2008 had many so-called swing states in which the election results were too close to predict accurately before election day and the margin for victory was narrow. Because of the close relationship between demographics and voting tendencies [2, 5–7, 13, 14, 20], this article examines the relationship between demographically similar counties and their voting tendencies for the 2008 U.S. presidential election in the three swing states with the most electoral votes (Florida, Pennsylvania, and Ohio). Emergent self-organizing maps (ESOMs) and k-means clustering were used to cluster demographic data provided by the 2010 U.S. Census, thereby identifying geographic regions (in this case, counties) that have similar demographics. After clustering, the voting results for the 2008 United States presidential election were examined within each each cluster of demographically similar counties. Sometimes, demographically similar counties in the same cluster voted for different candidates in the 2008 presidential election. The demographic clusters from ESOMs with mixed voting outcomes were examined closely, and it is suggested that a political party may be able to improve its chances for winning future elections

N. Mana, F. Schwenker, and E. Trentin (Eds.): ANNPR 2012, LNAI 7477, pp. 201–212, 2012.

by applying strategies that worked for one county in a cluster with mixed voting outcomes to other counties in the same cluster. The results obtained using ESOMs and k-means clustering to cluster census data for each of these three swing states are compared, and variation of information distance calculations were used to measure the dissimilarity between ESOM and k-means results [15]. ESOMs and k-means clustering were chosen to highlight the differences between ESOMs, which have thousands of weights (or neurons), and k-means clustering, which is believed to yield results similar to ordinary SOMs which have only tens or hundreds of weights [23].

2   Background

Self organizing maps were created by Finnish professor Teuvo Kohonen in the 1960s. A comprehensive mathematical description of Kohonen's work can be found in his book *Self-Organizing Maps* [10]. A key feature of a SOM is that it produces a low-dimensional picture (usually two- or three-dimensional) of a high-dimensional data set in such a way that points near each other in the low-dimensional picture come from points that are near each other in the high-dimensional data set. This dimension reduction feature of a SOM has been very useful to researchers interested in visualizing clusters in high-dimensional data sets.

SOMs have been used for a wide variety of applications, such as bioinformatics, health care, finance, language processing, document analysis, and image processing [12]. In a paper closely related to this one, Niemelä and Honkela used SOMs to explore the relationship between four socio-economic factors (cost of living, unemployment, gross domestic product, and total consumption) for the entire country of Finland between 1954 and 2003 and parliamentary election results that involved nine political parties during that time period [18]. In contrast, this study uses ESOMs to cluster counties in three states of the United states based on 51 different socio-economic factors measured in the 2010 Census and how they are related to the 2008 presidential election results between essentially two political parties (Democrats and Republicans). In work similar in spirit to this, Kaski and Kohonen used SOMs to study the socio-economic status of the countries in the world based on World-Bank data [9]. Tuia, et al., have used a SOM for the clustering of urban municipalities in Switzerland depending on their socio-economic profile [22]. The mechanism for self-organization in a SOM has even been used in a non-computational way as a metaphor to explain the patterns in electoral processes [16, 17]. The second author has used SOMs to identify benchmark universities on the basis of student assessment of university websites [4].

3   Methods

This section explains the data acquisition and preprocessing, how self-organizing maps were used to cluster the data, and how k-means clustering was used to cluster the data.

3.1   Data Acquisition and Preprocessing

The states of Florida, Pennsylvania, and Ohio were chosen for analysis because they are states in which the 2008 U.S. presidential election was close [3], they have the highest number of votes in the electoral college among all swing states (see Table 1), and they are projected to be states won by very narrow margins in the 2012 U.S. presidential election [1].

Table 1. Electoral college votes for three states in the 2008 and 2012 U.S. presidential elections

	Electoral College Votes		
Year	Florida	Pennsylvania	Ohio
2008	27	21	20
2012	29	20	18

Demographic data for all of the counties in these three swing states were obtained from the U.S. Census Bureau website for the year 2010 census [24]. The census data includes population demographics by age, ethnicity or race, education level, housing data, income data, employment data, trade data, government spending data, and land area data. A complete list of all 51 census data variables can be obtained from the Census Bureau website [25]. One data set was created for each of the three swing states from the 2010 census data, and the observations were the counties in the state and the variables were the 51 census data variables. Since the variables were not uniform in scale, each variable was normalized using a z-transform to make it scale invariant.

3.2   Clustering by Emergent Self-Organizing Maps

Emergent self organizing maps (ESOMs) are SOMs with several thousand weights (i.e., neurons), whereas ordinary SOMs have tens or hundreds of weights. ESOMs have been shown to be significantly different from ordinary SOMs, and ordinary SOMs can yield results similar to k-means clustering [23]. Thus, ESOMs were chosen for comparison to k-means clustering to determine whether ESOMs yield different clusterings than k-means does. The authors created ESOMs using the Databionic ESOM Analyzer [23], which uses the standard SOM algorithm for weight updates [10, 11]. The ESOMs created had 4,100 weights that were distributed on 50×82 toroidal grids in order to avoid error effects that occur near the edges of rectangular maps [23].

Distances between observations and weights in ESOMs were calculated using Pearson distance $1 - \rho$, where ρ is the Pearson correlation between an observation and a weight. Correlation was chosen for distance measurements instead of Euclidean distance because Euclidean distance may give undue influence to one particular variable. For example, the 2010 population of Philadelphia county is disproportional to the other counties in Pennsylvania, which is reflected by the fact that the z-score for Philadelphia county is 4.98 for the 2010 population variable. If Euclidean distance had been chosen for distance measurements, then very populous counties such as Philadelphia county would be separated by a sizable amount from counties with an average population, which would have a z-score near 0; hence, correlation distance was chosen to mitigate such effects. Also, using correlation tended to produce many clusters of small to medium size, whereas using Euclidean distance produced a few very large clusters and a smattering of very small clusters.

The topological ordering provided by the grid of an ESOM ensures that similar data points will be displayed in contiguous locations. However, by insisting on contiguous locations, the ESOM display suppresses the variation in the degrees of dissimilarity among the weights [21]. For this reason, a unified distance matrix, or U-matrix, was used to show inter-cluster distance in the ESOMs. The U-matrix was used to generate a contour map that visually separated clusters in the ESOMs.

The ESOMs were constructed using the following parameters. The number of training epochs was 30. The learning rate started at 0.95 and decreased linearly to 0.01. A two-dimensional output map grid on a torus was used with Euclidean distance on the grid. The other parameters used in the Databionic ESOM were the default values: online training was used, k for k-batch was 0.15, the initial map size was 10%, the ending epoch for good initialization was the 15th epoch, best matches were found using the standard search and a radius of 8, weights were initialized by a Gaussian distribution, the radius started at 24 and decreased linearly to 1, the neighborhood kernel function was a Gaussian, and the data patterns were permuted.

The ESOMs were configured to label each point with its county name and also color it according to how it voted in the 2008 U.S. presidential election. Voting results for the 2008 U.S. Presidential election were obtained from CNN [3]. After the Databionic ESOM software finished clustering counties in a state, some clusters of counties with similar demographics were then analyzed for how they voted in the 2008 U.S. Presidential election.

3.3   Clustering by k-Means

Clustering by k-means was chosen so that it could be compared to ESOM clustering. For a thorough explanation of k-means clustering, please see [8]. The k-means clustering method was used to divide the counties into $k = 30$ clusters. The choice of $k = 30$ clusters was used to make the cluster sizes for ESOMs and for k-means approximately the same. The k-means clustering was performed by the Lloyd-Forgy algorithm as implemented by the Kmeans function in the amap package of the statistics software R [19]. The initial means (or weight vectors) were chosen at random from the data, and a maximum of 100 iterations were allowed. As with the ESOMs, for k-means clustering the variables were normalized using a z-transform and Pearson distance was used as the metric.

3.4   Comparison of Clusterings Using Variation of Information

To measure the dissimilarity between ESOM and k-means clusterings, the variation of information distance was calculated between the ESOM clusterings and the k-means clusterings ($k = 30$). For more information on the variation of information distance metric, please see [15].

4   Results

This section presents the results obtained clustering the census data by ESOMs and k-means clustering. Also, clustering results for ESOMs and k-means are compared using the variation of information distance metric.

4.1   Results for ESOMs

The ESOMs for clustering counties in Florida, Pennsylvania, and Ohio are in Figures 1-3. These ESOMs use U-matrix maps to visualize the correlation between counties in these high-dimensional data sets, thereby creating a topographical map that shows mountain ranges for cluster boundaries, which are indicated by darker shading [23]. These ESOMs are toroidal maps, i.e., the top and bottom edges are identified, and the left and right edges are also identified. This means, for example, that Citrus, Charlotte, Sarasota, and Indian River counties belong to the same cluster in Figure 1, and that Ashland, Mercer, and Tuscarawas counties belong to the same cluster in Figure 3.

The ESOM for Florida in Figure 1 shows that, in general, counties in Florida clustered together by demographics tended to vote for the same presidential candidate in 2008. One notable exception to this was the cluster with Duval, Hillsborough, and Orange counties, which contain the major cities of Jacksonville, Tampa Bay, and Orlando, respectively. How the counties in this cluster voted in the 2008 U.S. presidential election is given in Table 2. The margins for victory in two of these three counties were very narrow. These three counties have very similar demographics since they're in the same cluster in the ESOM. Since political strategies often target particular demographics, strategies that worked in one county for a particular demographic should also work in another county that has similar demographics. That is, political parties could use their winning strategies from one of these three counties in the 2008 election to help them win in the other counties in this cluster in future elections.

The ESOM in Figure 2 shows that the counties in Pennsylvania clustered together by demographics also tended to vote for the same presidential candidate in 2008. The cluster containing Berks, Lancaster, Lehigh, and York counties contains four geographically contiguous counties with 13.5% of the population of the state. Two of these four counties voted for Obama in 2008 while the other two voted for McCain, as shown in Table 2. Since the 2012 U.S. presidential election is projected to be very close in Pennsylvania [1], political parties should consider applying winning strategies from one county in this cluster to all of the counties

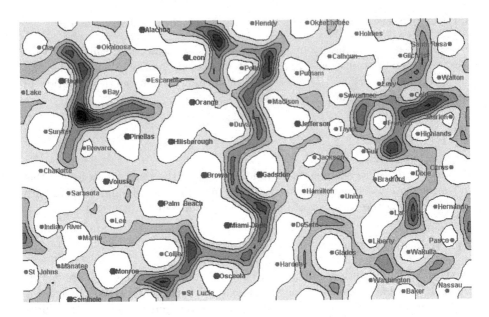

Fig. 1. ESOM clustering of Florida counties by year 2010 U.S. Census data on a toroidal map. Larger blue dots indicate counties that voted for Obama in 2008, while smaller red dots indicate counties that voted for McCain in 2008.

Table 2. (Left) One cluster of counties from the Florida ESOM and how the counties in this cluster voted in the 2008 U.S. presidential election. (Right) One cluster of counties from the Pennsylvania ESOM and how the counties in this cluster voted in the 2008 U.S. presidential election.

| Florida Voting (2008) | | | Pennsylvania Voting (2008) | | |
County	McCain	Obama	County	McCain	Obama
Duval	51%	49%	Berks	45%	54%
Hillsborough	46%	53%	Lancaster	55%	44%
Orange	41%	59%	Lehigh	42%	57%
			York	56%	43%

in this cluster to try to swing the vote in their favor. Also from the ESOM in Figure 2, there is a cluster that contains Cumberland and Chester counties, which are near the major cities of Harrisburg and Philadelphia, respectively, voted differently in 2008, and were decided by narrow margins. Since these two counties are closest on the ESOM, they have very similar demographics, which suggests that demographically based political strategies that were successful in one county may have a positive effect in the other county.

As was the case with Florida and Pennsylvania, the ESOM in Figure 3 shows that counties in Ohio clustered together by demographics also tended to vote for

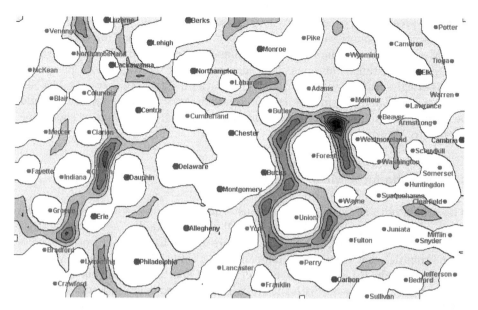

Fig. 2. ESOM clustering of Pennsylvania counties by year 2010 U.S. Census data on a toroidal map. Larger blue dots indicate counties that voted for Obama in 2008, while smaller red dots indicate counties that voted for McCain in 2008.

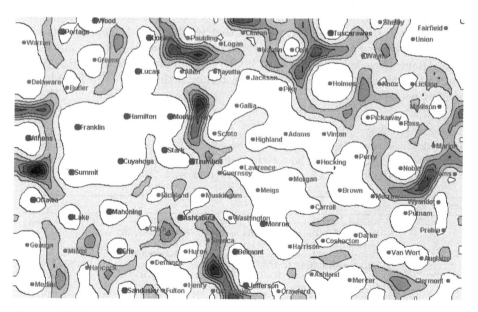

Fig. 3. ESOM clustering of Ohio counties by year 2010 U.S. Census data on a toroidal map. Larger blue dots indicate counties that voted for Obama in 2008, while smaller red dots indicate counties that voted for McCain in 2008.

the same presidential candidate in 2008. The cluster containing Ashland, Mercer, and Tuscarawas counties contains one county, Tuscarawas, that voted for Obama over McCain by a margin of 50% to 48%, and two other counties that voted for McCain by substantial margins. This suggests that if the Republican party had used strategies from Ashland and Mercer counties in Tuscarawas county, McCain may have been able to win Tuscarawas county in 2008. A similar argument could be made that political strategies from Meigs or Harrison counties should be applied to Monroe county.

4.2   Results for k-Means Clustering

The results obtained using k-means clustering to create $k = 30$ clusters for the counties in Florida, Pennsylvania, and Ohio are in Tables 3-5. For the these three states, the number of ESOM clusters which have mixed voting results (some counties that voted for Obama and other counties that voted for McCain) are 2 for Florida, 2 for Pennsylvania, and 3 for Ohio. In contrast, the number of k-means clusters ($k = 30$) with such mixed voting results are 9, 4, and 7, respectively. Thus, assuming there is a very close relationship between demographics and presidential voting, the fact that the ESOMs have fewer clusters with mixed voting results may indicate that ESOMs perform better at constructing homogeneous clusters than k-means clustering. Detailed qualitative analysis of the k-means clusterings also reveals results such as that Okaloosa and Miami-Dade counties are in the same cluster, which is somewhat surprising since Okaloosa is considered to be one of the most conservative (Republican) counties in Florida while Miami-Dade is one of the most liberal (Democrat). This may be evidence that the choice of $k = 30$ for k-means clustering of the counties in Florida was perhaps not optimal.

Table 3. Clustering of counties in Florida by year 2010 U.S. Census data via k-means (with $k = 30$) together with how each county voted in the 2008 U.S. presidential election (D = Democrat = Obama, R = Republican = McCain).

Cluster size	Clusters
6	{(R) Franklin, (R) Hernando, (D) Jefferson, (R) Lake, (R) St. Johns, (R) Santa Rosa}
5	{(R) Hardee, (D) Hillsborough, (R) Sarasota, (R) Wakulla, (R) Washington}
4	{(R) Collier, (R) Dixie, (D) Flagler, (R) Indian River}
3	{(R) Holmes, (D) Miami-Dade, (R) Okaloosa}, {(R) Calhoun, (R) Gulf, (R) St. Lucie}, {(R) Brevard, (R) Citrus, (R) Glades}, {(R) Lafayette, (R) Liberty, (R) Martin}, {(R) Duval, (R) Sumter, (R) Taylor}, {(R) Bradford, (R) Gilchrist, (R) Jackson}, {(R) Hamilton, (D) Leon, (R) Nassau}, {(R) Clay, (R) Hendry, (R) Madison}
2	{(R) Manatee, (R) Walton}, {(R) Marion, (D) Pinellas}, {(D) Monroe, (R) Polk}, {(R) Charlotte, (R) Baker}, {(D) Orange, (D) Volusia}, {(D) Palm Beach, (D) Seminole}, {(D) Gadsen, (R) Pasco}, {(R) Highlands, (R) Union}, {(R) Lee, (D) Osceola}
1	{(R) Columbia}, {(R) Bay}, {(D) Broward}, {(R) Okeechobee}, {(R) Putnam}, {(R) Suwanee}, {(R) Escambia}, {(R) DeSoto}, {(R) Alachua}, {(R) Levy}

Table 4. Clustering of counties in Pennsylvania by year 2010 U.S. Census data via k-means (with $k = 30$) together with how each county voted in the 2008 U.S. presidential election (D = Democrat = Obama, R = Republican = McCain).

Cluster size	Clusters
11	{(R) Bedford, (R) Bradford, (R) Clearfield, (R) Crawford, (R) Jefferson, (R) Lycoming, (R) McKean, (R) Potter, (R) Tioga, (R) Venango, (R) Warren}
5	{(R) Beaver, (R) Blair, (D) Cambria, (R) Lawrence, (R) Mercer}, {(R) Clarion, (R) Clinton, (R) Columbia, (R) Greene, (R) Indiana}, {(R) Armstrong, (R) Huntingdon, (R) Schuylkill, (R) Somerset, (R) Susquehanna}
3	{(R) Cameron, (D) Elk, (R) Montour}
2	{(D) Monroe, (R) Pike}, {(R) Sullivan, (R) Wayne}, {(R) Franklin, (R) Lebanon}, {(R) Butler, (R) Washington}, {(R) Adams, (R) Wyoming}, {(D) Lackawanna, (D) Luzerne}, {(R) Forest, (R) Union}, {(D) Bucks, (D) Montgomery}, {(R) Mifflin, (R) Northumberland}, {(D) Carbon, (R) Fulton}, {(D) Dauphin, (D) Erie}, {(R) Juniata, (R) Snyder}, {(D) Allegheny, (D) Philadelphia}
1	{(R) Cumberland}, {(D) Chester}, {(D) Northampton}, {(D) Lehigh}, {(D) Berks}, {(D) Centre}, {(R) Fayette}, {(R) Perry}, {(R) Lancaster}, {(R) York}, {(D) Delaware}, {(R) Westmoreland}

Table 5. Clustering of counties in Ohio by year 2010 U.S. Census data via k-means (with $k = 30$) together with how each county voted in the 2008 U.S. presidential election (D = Democrat = Obama, R = Republican = McCain).

Cluster size	Clusters
7	{(R) Carroll, (R) Guernsey, (R) Highland, (R) Lawrence, (R) Meigs, (R) Morgan, (R) Washington}, {(R) Adams, (R) Fayette, (R) Gallia, (R) Jackson, (R) Pike, (R) Scioto, (R) Vinton}
6	{(D) Belmont, (R) Coshocton, (R) Crawford, (R) Harrison, (D) Jefferson, (D) Monroe}, {(R) Brown, (R) Darke, (R) Hocking, (R) Knox, (R) Morrow, (R) Perry}
5	{(D) Ottawa, (R) Preble, (R) Shelby, (R) Williams, (R) Wyandot}, {(R) Greene, (R) Hancock, (R) Miami, (D) Portage, (D) Wood}, {(D) Defiance, (R) Fulton, (R) Henry, (R) Huron, (D) Sandusky}, {(D) Cuyahoga, (D) Hamilton, (D) Montgomery, (D) Stark, (D) Summit}
4	{(R) Ashland, (R) Auglaize, (R) Mercer, (R) Putnam}, {(R) Columbiana; (R) Marion, (R) Noble, (R) Pickaway}
3	{(R) Clark, (D) Erie, (D) Mahoning}, {(R) Paulding, (R) Seneca, (R) Van Wert}, {(R) Delaware, (R) Fairfield, (R) Warren}, {(R) Allen, (R) Richland, (D) Trumbull}
2	{(R) Geauga, (R) Medina}, {(R) Madison, (R) Ross}, {(D) Lorain, (D) Lucas}, {(R) Ashtabula, (D) Muskingum}, {(R) Clinton, (R) Hardin}, {(R) Butler, (D) Franklin}
1	{(R) Wayne}, {(R) Holmes}, {(D) Athens}, {(R) Champaign}, {(R) Logan}, {(D) Tuscarawas}, {(R) Clermont}, {(R) Union}, {(D) Lake}, {(R) Licking}

4.3   Comparison of Clusterings Using Variation of Information

To compare the ESOM and k-means clusterings for each state, variation of information distance (with a base 2 logarithm) was used [15]. The variation of information results are shown in Table 6, and they were produced using a Matlab / Octave function written by the first author. Since ESOM clustering used a U-matrix, its clustering results show somewhat continuous variation between clusters. In contrast, the variation between k-means clusters is discrete. To make the ESOM clustering results discrete for the purpose of using variation of

Table 6. Variation of information (VI) distances between ESOM clustering and k-means clustering ($k = 30$)

	Florida	Pennsylvania	Ohio
Actual VI distance	1.9856	1.6366	1.6585
Max possible VI distance	$\log_2(67)$	$\log_2(67)$	$\log_2(88)$
(Actual VI) / (Max possible VI)	0.32733	0.26979	0.25676

information to compare it with k-means clustering, counties in an ESOM were considered to be in distinct clusters whenever a level curve on the topographical map would have to be crossed, and in the same cluster otherwise. Since none of the variation of information distances in Table 6 are close to zero, the ESOM and k-means clusterings are measurably different. The results in Table 6 support the claim made by Ultsch and Moerchen [23] that ESOM clusterings are different from k-means clusterings, which are believed to be very similar to ordinary SOM clusterings.

5   Discussion

The article by Nate Silver [20] emphasizes that every individual is a member of many demographic categories and "the voting tendencies associated with those categories often point in different, or even conflicting, directions." Despite this, political parties have tried to target individual demographic categories instead of incorporating strategies that address several demographic categories simultaneously. Using an ESOM to cluster geographic regions by a broadly defined set of demographic data could help political parties identify possibly non-contiguous geographic regions that could benefit from demographically targeted political strategies. Additionally, using an ESOM to cluster groups of people across many demographic categories at once, and then pairing this information with voting tendencies could help produce a clearer picture of how demographics are related to voting tendencies. This information could play a vital and pivotal role for political parties trying to win elections in very close races.

Future directions for research in this area may include sensitivity analysis of ESOMs obtained by looking at how an ESOM changes when one of its input variables changes, comparing the results of ESOMs to other exploratory clustering methods by using variation of information or some other method of comparing clusterings, or using ESOMs in a similar manner on a smaller, more narrowly defined, set of demographic variables to try to produce ESOMs having even fewer clusters with non-homogeneous voting outcomes.

References

1. 2012 Presidential Election Interactive Map and History of the Electoral College, http://www.270towin.com/ (retrieved May 5, 2012)
2. Ansolabehere, S., Persily, N., Stewart, C.: Race, Region, and Vote Choice in the 2008 Election: Implications for the Future of the Voting Rights Act. Harvard Law Review 123 (2010); Columbia Public Law Research Paper No. 09-211; MIT Political Science Department Research Paper No. 2011-1. Available at SSRN: http://ssrn.com/abstract=1462363 (retrieved May 1, 2012)
3. CNN. County Results - Election Center 2008 - Elections & Politics from CNN.com, http://www.cnn.com/ELECTION/2008/results/county/ (retrieved May 1, 2012)
4. Cooper, C., Burns, A.: Kohonen Self-Organizing Feature Maps as a Means to Benchmark College and University Websites. Journal of Science Education and Technology 16(3), 203–211 (2007)
5. Frey, W.: Battling Battlegrounds. American Demographics (September 24–26, 2004)
6. Gelman, A., Kenworthy, L., Su, Y.: Income Inequality and Partisan Voting in the United States. Social Science Quarterly, Special Issue: Inequality and Poverty: American and International Perspectives 91(5), 1203–1219 (2010)
7. Gimpel, J., Dyck, J., Shaw, D.: Registrants, Voters, and Turnout Variability Across Neighborhoods. Political Behavior 26(4), 343–375 (2004)
8. Hartigan, J.: Clustering Algorithms, pp. 1–351. Wiley, New York (1975)
9. Kaski, S., Kohonen, T.: Exploratory Data Analysis by the Self-Organizing Map: Structures of Welfare and Poverty in the World. In: Refenes, A., Abu-Mostafa, Y., Moody, J., Weigend, A. (eds.) Neural Networks in Financial Engineering, pp. 498–507. World Scientific, Singapore (1996)
10. Kohonen, T.: Self-Organizing Maps, 3rd edn., pp. 1–521. Springer, Berlin (2000)
11. Kohonen, T.: The Self-Organizing Map. Proceedings of the IEEE 78(9), 1464–1480 (1990)
12. Laaksonen, J., Honkela, T. (eds.): WSOM 2011. LNCS, vol. 6731, pp. 1–380. Springer, Heidelberg (2011)
13. Lesthaeghe, R., Niedert, L.: US Presidential Elections and the Spatial Pattern of the American Second Demographic Transition. Population and Development Review 35(2), 391–400 (2009)
14. Lopez, M.: Dissecting the 2008 Electorate: Most Diverse in U.S. History - Pew Research Center, http://pewresearch.org/pubs/1209/racial-ethnic-voters-presidential-election (retrieved May 1, 2012)
15. Meilă, M.: Comparing Clusterings by the Variation of Information. In: Schölkopf, B., Warmuth, M.K. (eds.) COLT/Kernel 2003. LNCS (LNAI), vol. 2777, pp. 173–187. Springer, Heidelberg (2003)
16. Neme, A., Hernández, S., Neme, O.: An Electoral Preferences Model Based on Self-Organizing Maps. Journal of Computational Science 2, 345–352 (2011)
17. Neme, A., Hernández, S., Neme, O.: Self Organizing Maps as Models of Social Processes: The Case of Electoral Preferences. In: Laaksonen, J., Honkela, T. (eds.) WSOM 2011. LNCS, vol. 6731, pp. 51–60. Springer, Heidelberg (2011)
18. Niemelä, P., Honkela, T.: Analysis of Parliamentary Election Results and Socio-Economic Situation Using Self-Organizing Map. In: Príncipe, J.C., Miikkulainen, R. (eds.) WSOM 2009. LNCS, vol. 5629, pp. 209–218. Springer, Heidelberg (2009)
19. The R project for statistical computing, http://www.r-project.org/ (retrieved June 20, 2012)

20. Silver, N.: In Politics, Demographics Are Not Destiny - NYTimes.com, http://fivethirtyeight.blogs.nytimes.com/2011/03/01/in-politics-demographics-are-not-destiny/ (retrieved May 1, 2012)
21. Trosset, M.: Representing Clusters: K-Means Clustering, Self-Organizing Maps, and Multidimensional Scaling, Technical Report 08-03, Department of Statistics, Indiana University, Bloomington, IN (2008)
22. Tuia, D., Kaiser, C., Da Cunha, A., Kanevski, M.: Socio-economic Data Analysis with Scan Statistics and Self-organizing Maps. In: Gervasi, O., Murgante, B., Laganà, A., Taniar, D., Mun, Y., Gavrilova, M.L. (eds.) ICCSA 2008, Part I. LNCS, vol. 5072, pp. 52–64. Springer, Heidelberg (2008)
23. Ultsch, A., Moerchen, F.: ESOM-Maps: tools for clustering, visualization, and classification with Emergent SOM, Technical Report Dept. of Mathematics and Computer Science, University of Marburg, Germany, No. 46 (2005)
24. United States Census Bureau. Download QuickFacts from the US Census Bureau, http://quickfacts.census.gov/qfd/download_data.html (retrieved May 1, 2012)
25. United States Census Bureau. Dictionary of census data, http://quickfacts.census.gov/qfd/download/DataDict.txt (retrieved May 1, 2012)

Grayscale Images and RGB Video:
Compression by Morphological Neural Network

Osvaldo de Souza[1], Paulo César Cortez[1], and Francisco A.T.F. da Silva[2]

[1] Federal University of Ceará, DETI, Fortaleza, Brazil
osvaldo@ufc.br, cortez@lesc.ufc.br
[2] National Institute For Space Research, ROEN, Eusébio, Brazil
tavares@roen.inpe.br

Abstract. This paper investigates image and RGB video compression by a supervised morphological neural network. This network was originally designed to compress grayscale image and was then extended to RGB video. It supports two kinds of thresholds: a pixel-component threshold and pixel-error counting threshold. The activation function is based on an adaptive morphological neuron, which produces suitable compression rates even when working with three color channels simultaneously. Both intra-frame and inter-frame compression approaches are implemented. The PSNR level indicates that the compressed video is compliant with the desired quality levels. Our results are compared to those obtained with commonly used image and video compression methods. Network application results are presented for grayscale images and RGB video with a 352×288 pixel size.

Keywords: Supervised Morphological Neural Network, RGB Video Compression, and Image Compression.

1   Introduction

The loss of data is common in a variety of image and video compression techniques, and such losses generally occur in parts of the information (redundancy data) that are not noticed by human eyes. Numerous compression algorithms utilize common techniques such as *"color space sampling"* and *"redundancy reduction"* [1]. The color space sampling technique is used when it is necessary to reduce the amount of data needed for the representation (coding) of an image. In the redundancy reduction technique, compression can be obtained by eliminating the redundancies that appear in a specific frame (intra-frame) or in a sequence of frames (inter-frame) in a video stream. Several studies have investigated the use of artificial neural networks (ANN) in image and video compression [1]. Some researchers [2] investigated image compression and reconstruction using a radial basis function (RBF) network, while others [3] proposed a technique called a "point process" that used a combination of motion estimation, compression, and temporal frame sub-sampling with a random neural network (RNN). In [4], the authors discussed various ANN architectures for

N. Mana, F. Schwenker, and E. Trentin (Eds.): ANNPR 2012, LNAI 7477, pp. 213–224, 2012.

image compression and presented the results for a back-propagation network (BPN), hierarchical back-propagation network (HBPN), and adaptive back-propagation network (ABPN). In [5], they used a self-organizing map (SOM) network to reduce the number of pixels in each frame of a video sequence. After this modification, each frame was stored using a Hopfield neural network as a form of video codification. In [6], they used the growing neural gas (GNG) learning method, another approach based on a SOM network, in an incremental training algorithm. In [7], the authors presented the details of an approach in which a neural network is used to determine the best ratio for discrete cosine transform (DCT) compression. Although there have been many works related to image and video compression, the use of supervised morphological neural networks (SMNNs) in this context has not been extensively investigated thus far. Therefore, in this paper, we investigate the extension of an SMNN, which was originally designed to compress grayscale images, by applying it to the compression of RGB video.

We organize the remaining sections of this paper as follows. We first provide a brief review of the morphological operators involved in the design of the adaptive morphological neuron. Second, we introduce the SMNN for grayscale image compression and then extend its application to RGB video compression. Third, we present the image and video compression results.

2   Brief Review of Morphological Operators

The morphological operators presented in this section were defined in [8] and briefly in [9], while the researchers in [10] proposed a morphological approach for template matching.

Definition 1. Let E be a non-void set in $\mathbf{z}^2$ and l be an integer number between 0 and n. K_n^E in K_l^E, denoted by $\delta_l\ \varepsilon_l\ \varepsilon_l^a\ \delta_l^a$, are operators defined as dilation, erosion, anti-dilation, and anti-erosion, respectively. Formal definitions for these operators are given in [8] and [10].

Definition 2. Let a window W be a non-void subset of $\mathbb{Z}^2$. An individual element of W is denoted by w, according to:

$$w \in W \mid W \subset \mathbb{Z}^2. \tag{1}$$

Definition 3. Let $W \subset \mathbb{Z}^2$ be a window and D and E be two non-void subsets of $\mathbb{Z}^2$, such that $D = E \oplus W$, and let l be an integer number between 0 and m. The symbol $\oplus$ refers to Minkowski addition. We denote by ε_l^i and δ_l^{ai} the operators from K_m^D to K_1^E defined in [10].

Definition 4. Let $f \in K_n^W$ and c_1 and c_2 $(c_1 < c_2)$ be two integer constants. We define the following functions from W to K_n:

$$(f_w^-)(x) = max\{0, min(n, f_w(x) + c_1)\}, \tag{2}$$

$$(f_w^+)(x) = max\{0, min(n, f_w(x) + c_2)\}, \tag{3}$$

where $x \in W$. The values of c_1 and c_2 are calculated according to de following equation.

$$d\mu = \mu(g) - \mu(f) \qquad c_1 = d\mu - \frac{F}{2} \qquad c_2 = d\mu + \frac{F}{2} \tag{4}$$

where the length of F is the interval $[c_1, c_2]$, centered at $d\mu$.

Definition 5. Let ε_t^i and δ_t^{ai} be the operators from K_m^D to K_1^E given by definition 1, and let f_w^- and f_w^+ be the functions defined by (2) and (3), respectively. λ^i is an operator from K_m^D to K_1^E defined by:

$$\lambda^i = \varepsilon_{f_w^-(w_i)}^i \wedge \delta_{f_w^+(w_i)}^{ai}, \tag{5}$$

where W is a window; $m = \#W; i = 1, \ldots, m$.

Definition 6. The pattern matching operator from K_m^D to K_m^E is:

$$\phi = \sum_{i=1,\ldots,n} \lambda^i \tag{6}$$

The operator ϕ represents the intersections between erosions and anti-dilations with the tolerances introduced by equations (2) and (3), which are controlled by the value of F. Therefore, such operations start to behave as morphological operators with gray level tolerance. Observe that the operations in equation (6) result in adaptive pattern matching.

Definition 7. Let $t \in K_n$ be a threshold. The operator $\overset{\bullet}{\Psi_t}$ from K_n^E to K_1, which localizes a concentration of gray levels above or equal to t, according to [11] is defined as:

$$\overset{\bullet}{\Psi_t} \circ \phi = \begin{cases} 1, & if \; \exists \, x \in E, \quad f(x) \geq t \\ 0, & otherwise \end{cases} \tag{7}$$

This threshold operator is a morphological filter, which is useful for adaptive pattern detection, and it is a key component in the activation function of the morphological neuron used in this work, as discussed later. The equations and definitions presented in this section are first applicable to gray scale images. Thus, it is important to note that because the color components in RGB schema can be represented in a range of values between 0 to 255, all of the definitions and proofs available in [8-10] and [12] that were developed for gray scale images are suitable for processing color images, since we consider only one component of the color at a time. This strategy is adopted in this work, and we refer to the values between 0 to m of the color component, as the "color variation of the component" (CVC), where $0 < m < 255$.

3   SMNN for Image Compression

The activation potential and activation function of SMNN for image compression are based on equations (6) and (7), and they are defined according to:

$$v_k = \phi,$$
(8)

and

$$y_k = \Psi_t^{\bullet} \circ \phi.$$
(9)

During the supervised training, the weights first decay in order to accelerate the weight adjustment process, according to equation (10):

$$W_{k_n} = W_{k_{n-1}} \left(1 - \left(\varepsilon * \Delta_{m_n}\right)\right).$$
(10)

Where $0 < \varepsilon < 1$, W_{k_n} refers to the weight of neuron k in iteration n, and Δ_m is defined according to

$$\Delta_m = 1 - \left(\left(\sum_{i=1}^{MN} A\right) \middle/ \Delta max_m\right).$$
(11)

Then the weights are adjusted according to

$$W_{k_n} = W_{k_{n-1}} + \beta \times \left(D_k - W_{k_{n-1}}\right),$$
(12)

where D_k is the desired value, and β Is an array, defined by

$$\beta = \mu\rho.$$
(13)

Note that μ is the learning constant and ρ is an array defined by the following equations

$$A = \left\{\varepsilon_{f_w^-(w_i)}^i \wedge \delta_{f_w^+(w_i)}^{ai}\right\},$$
(14)

where A is an array with dimensions K_1^W.

$$V_e = A^C,$$
(15)

where A^C is the complement of A,

$$\Delta max_m = M * N,$$
(16)

where M and N are the dimensions of A^C, and finally we have

$$\rho = \Delta_m V_e.$$
(17)

The activation function for the morphological neuron in the auxiliary layer is:

$$\delta_{D_k}^*(y_k) = \begin{cases} f_{D_k}, & if \ y_k = 1 \\ 0_D, & otherwise \end{cases}$$
(18)

Note that equation (18) is a morphological dilation. Figure 1 presents the architecture for this SMNN; observe that the network is composed of an input layer, an output layer, and a hidden layer with its auxiliary layer. The input layer receives the patterns to be learned; in this case, "patterns" refers to the data of the image to be compressed.

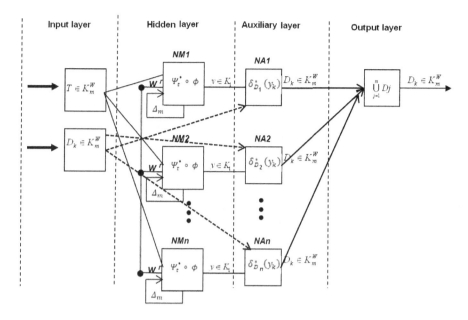

Fig. 1. Architecture of a SMNN

In this architecture, the T patterns presented to the input layer must belong to the K_m^W domain; the D_k value refers to the desired value, which is mandatory for error correction in the learning phase. The network's output is limited to the K_m^W domain. K_m^W refers to a sub-image of dimension W with a positive grayscale level between [0, m]. $y \in K_1$ refers to the output of a morphological neuron of the second layer.

According to definition 4 and equations (2–4) SMNN allows the definition of a pixel-component threshold, in fact, the tolerance interval F. This interval defines the tolerance of the SMNN to deal with gray level variations between the neurons' weights and a pattern under processing by the network. In addition, SMNN also allows the definition of a pixel-error counting threshold. This threshold is responsible for restricting the neuron's activation. In the following sections, we extend a SMNN in order to make it capable of compressing images and RGB video.

3.1 Grayscale Image Compression

The image to be compressed is fragmented into a set of windows. Each element of this set is processed by all of the morphological neurons (MNs). The Winner neuron produces an output with value 1 (high), while all of the others produce outputs with value 0 (low). *Let $I \in K_m^{MN}$ be an* image of dimensions $M \times N$ and positive grayscale levels within [0, m], and W_i be a sub-image of I, such that the union of all of the sub-images reconstructs the original image I. The reconstruction of image I is defined by:

$$I = \bigcup_{1}^{s} w_i \tag{19}$$

One extension of SMNN is required for image compression: the auxiliary neuron is loaded with the sequential number of the corresponding MN from the second layer. The output of the MN is received by its corresponding outstar neuron, which when excited with a high input, outputs the value loaded during the training phase. Thus, the network effectively indicates the winner MN that has learned or recognized the input pattern (which in this particular application refers to a W_i window). This output must be preserved, mapping which W_i window a MN can reproduce. This is the key for decoding the compressed image; we use a mapping between a window and the neuron that has "learned" this window. In this way, the *nth* neuron's weights are used in order to reproduce the windows associated with it. This mapping is defined by

$$I^r = \{(k,n)_1, \dots, (k,n)_i \mid \{k \in K, n \in S\}\}, \tag{20}$$

where S is the set of w_i sub-images, regarding image I. The value of k means the *kth* MN and $(k,n)_i$ refers to the *ith* mapping between a MN and a sub-image (window). Figure 2 shows how an image or a component of the frame is processed by a SMNN.

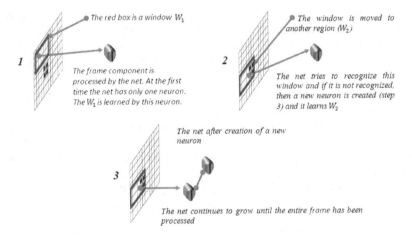

Fig. 2. Illustration of how image or frame is compressed by the SMNN

3.2   RGB Video Compression

For the compression of RGB video, each component of the stream is processed in an individual instance of SMNN. Consequently, at the end of the compression, we obtain three instances of SMNN. This process is depicted in Figure 3.

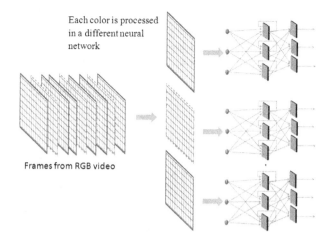

Fig. 3. RGB video compression by SMNN

The use of SMNN for video stream compression does not require any adaptation of its extended proposition for grayscale image compression. Each frame is split into a set of w_i sub-images. Then, for each set, we create a I^r map, according to equation (20). Finally, an entire component stream is encoded according to:

$$\bigcup_{i=1}^{N} I_i^r \, ,$$ (21)

where N is the length of the video stream measured in frames. In the next section we present the SMNN results related to the compression of grayscale images and RGB video.

4   Computer Simulations and Discussion

Table 1 presents the results for the compression of Figure 2(G), which is a grayscale image with 320X240 pixels and 8 bits per pixel (bpp), totaling 76.800 bytes without compression.

Table 1. Results obtained by applying 3 image compression methods to figure 2(G)

Image	Format	Image Size MxN	Size (bytes)	Compression ratio C_R	Fidelity Criteria eRMS
A	PNG	320x240	77.279	(0,99)	2.6
B	PNG	320x240	54.858	1.39	8.4
C	NMC	320x240	72.040	1.07	0.3
D	NMC	320x240	30.146	2.55	4.9
E	JPG	320x240	41.684	1.84	3.6
F	JPG	320x240	2.468	31.11	27.5

Measurements of $eRMS$ and C_R are always calculated in relation to image 2(G). In Table 1, JPG refers to the Joint Photographic Experts Group format, PNG refers to the portable network graphics and NMC refers to the neural morphological compression method, produced by SMNN. Each method was used to produced images with the highest and lowest compression levels possible. To evaluate the fidelity criteria, we use the root mean square error (eRMS) for an objective evaluation of the images in Figure 2(A-G).

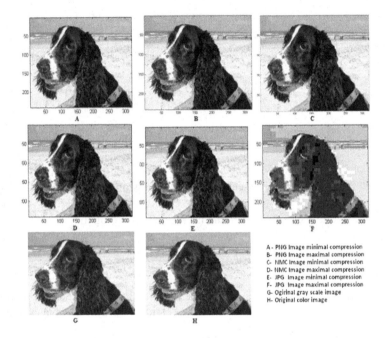

A - PNG Image minimal compression
B - PNG Image maximal compression
C - NMC Image minimal compression
D - NMC Image maximal compression
E - JPG Image minimal compression
F - JPG Image maximal compression
G - Ogirinal gray scale image
H - Original color image

Fig. 4. Dog Lisbela in different images formats obtained with 3 compression methods

The compression ratio estimation listed in Table 1 were obtained in accordance with equation (22), in which s_1 refers to original image size in bytes, s_2 refers to compressed image size in bytes, and $eRMS$ are defined according to equation (23).

$$C_{R=} \frac{s_1}{s_2} \tag{22}$$

$$e_{rms=} \left[\frac{1}{MN} \sum_{x=0}^{M-1} \sum_{y=0}^{N-1} [f_2(x,y) - f_1(x,y)]^2 \right]^{1/2} \tag{23}$$

Figure 5(A) depicts the results for compression with a window size = 3 and the following variations in the SMNN's parameters: T from 0.7 to 1.0, and F from 5 to 10. Note that in charts (A) and (B), the value of F was normalized. In chart (B), we can see the results for a window size = 3 and the same variation in the SMNN's parameters as seen in chart (A).

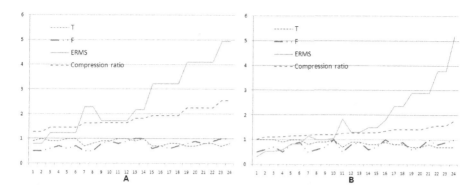

Fig. 5. Results for compression of grayscale image with variations of SMNN's parameters

For RGB video compression we used the "foreman" stream [13]. This test video was obtained by converting a CIF video to the RGB color space, sampling with 10 frames. The results are presented for 8×8 and 16×16 window sizes. In Figure 5 we can see samples of frame ten compressed with variations of SMNN's parameters.

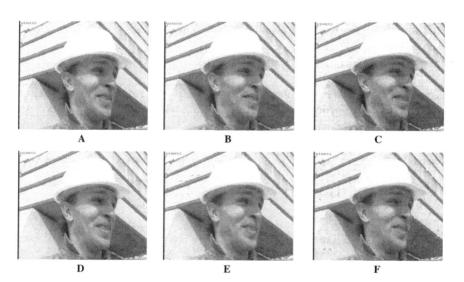

Fig. 6. Frame ten compressed with various parameters.

Table 2. Samples of results obtained by applying SMNN to 10th frame of *Foreman* video

Frame	Window Size MxN	F	t	Compression ratio C_R	Objective Quality $PSNR$ (db)
A	8×8	10	0.90	1.296	45.47
B	8×8	20	0.90	2.357	34.33
C	8×8	30	0.90	3.698	30.25
D	16×16	10	0.90	1.162	40.31
E	16×16	20	0.90	1.690	34.20
F	16×16	30	0.90	2.227	30.59

The foreman video was successfully compressed by a SMNN, as we can see in table 2, for a windows size 8×8, with a low tolerance to variations in the pixel values ($F = 10$) and a pixel-error threshold setting of $t=0.90$, resulting in a good compression ratio (3.69) and an acceptable PSNR level. Note that these results refer to the 10th frame only. To evaluate the fidelity criteria for the compressed images and video we utilize the peak signal-to-noise ratio (PSNR) for an objective evaluation according to equation:

$$PSNR = 10.\log_{10}\left(\frac{MAX_I^2}{MSE}\right) \qquad (24)$$

We extended the investigation of RGB video compression by compressing the first 100 frames of the Foreman video, and compared the results of SMNN with results from other well-known compression techniques. Table 3 shows these results. Figure 6 shows the PSNR and CR evolution, frame-by-frame, throughout the compression of the first 100 frames of the Foreman video, while Figure 7 shows the growth in the number of neurons during this compression.

In Table 3 the results for HEO-II refers to [14], KAMINSKY and JM9.5 refer to [15], and FS and ANEA refer to [16]. NMC1 refers to the results for SMNN with $t=0.6$, $F=5.0$, and a window with 8×8 pixels size, and NMC2 refers to results for SMNN with $t=0.8$, $F=20.0$, and a window with 4×4 pixels size.

Table 3. Numerical results obtained by applying SMNN to first 100 frames of *Foreman* video.

Technique	Requires complex pre-processing?	PSNR	Bit rate (bits/pixel)	Compression ratio
HEO-II-100F	(yes) H264/AVC	NA	0.660	1.51
KAMINSKY-100F	(yes) H264/AVC	35.85	0.0702	14.282
JM9.5-100F	(yes) H264/AVC	35.93	0.0705	14.182
FS	(yes) H264/AVC	36.33	0.384	2.60
ANEA	(yes) H264/AVC	36.29	0.543	1.84
NMC1	(none)	34.30	0.380	2.63
NMC2	(none)	27,93	0.104	9,59

In Figure 7 the sub-images refer to the number of elements in I_i^r, measured at the 100th frame (equation (21)).

As we can see in Tables 1 and 3, SMNN gave good results demonstrating that the network is capable of RGB video compression. Note that SMNN does not require pre-processing and all of results shown in this paper refer to the data without any secondary compression. Saving SMNN results to a hard-disk using trivial data compression can improve the final compression rates.

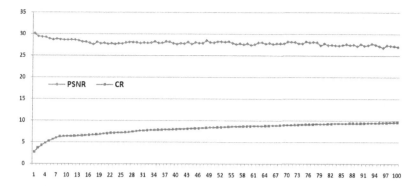

Fig. 7. Frame-by-frame evolution during compression of first 100 frames

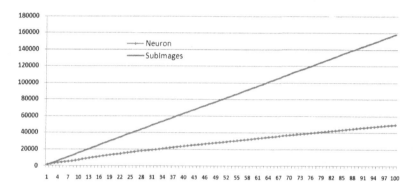

Fig. 8. Growth in the numbers of neurons during compression

5   Conclusion

This investigation and the detailed results for SMNN demonstrated that it is practical for RGB video and grayscale image compression and capable of producing results comparable to well-known methods. The reconstruction of the compressed image essentially occurs through data translation from neuron's weights to the respective *windows*, without requiring additional mathematical operations.

References

1. Winkler, S., van den Branden Lambrecht, C.J., Kunt, M.: Vision Models and Applications to Image and Video Processing, p. 209. Springer (2001)
2. Reddy, et al.: Image Compression and Reconstruction Using a New Approach by Artificial Neural Network. International Journal of Image Processing (IJIP) 6(2), 68–85 (2012)
3. Cramer, C., Gelenbe, E., Bakircloglu, H.: Low Bit-rate Video Compression with Neural Networks and Temporal Subsampling. Proceedings of the IEEE 84(10), 1529–1543 (1996)

4. Vaddella, R.P.V., Rama, K.: Artificial Neural Networks for Compression of Digital images: A Review. International Journal of Reviews in Computing, 75–82 (2010)
5. Singh, M.P., Arya, K.V., Sharma, K.: Video Compression Using Self-Organizing Map and Pattern Storage Using Hopfield Neural Network. In: International Conference on Industrial and Information Systems (ICIIS), December 28-31, pp. 272–278 (2009)
6. García-Rodríguez, J., Domínguez, E., Angelopoulou, A., Psarrou, A., Mora-Gimeno, F.J., Orts, S., García-Chamizo, J.M.: Video and Image Processing with Self-Organizing Neural Networks. In: Cabestany, J., Rojas, I., Joya, G. (eds.) IWANN 2011, Part II. LNCS, vol. 6692, pp. 98–104. Springer, Heidelberg (2011)
7. Khashman, A.: Neural Networks Arbitration for Optimum DCT Image Compression. In: IEEE Eurocon (2007)
8. Banon, G.J.F.: Characterization of Translation Invariant Elementary Morphological Operators Between Gray-level Images. INPE, São José dos Campos, SP, Brasil (1995)
9. Banon, G.J.F., Faria, S.D.: Morphological Approach for Template Matching. In: Brazilian Symposium on Computer Graphics and Image Processing Proceedings. IEEE Computer Society (1997)
10. Faria, S.D.: Uma abordagem morfológica para casamento de padrões, Master Tesis, National Institute for Space Research, INPE-6346-RDI/597 (1997)
11. Silva, F.A.F.S., Banon, G.J.F.: Rede morfológica não supervisionada (RMNS). In: IV Brazilian Conference on Neural Networks, pp. 400–405 (1999)
12. Banon, G.J.F., Barrera, J.: Decomposition of Mappings Between Complete Lattices by Mathematical Morphology – Part I: General Lattices. Signal Processing 30(3), 299–327 (1993)
13. Foreman, Video stream for tests, http://trace.eas.asu.edu/yuv/
14. Heo, J., Ho, Y.-S.: Efficient Differential Pixel Value Coding in CABAC for H.264/AVC Lossless Video Compression. Circuits, Systems and Signal Processing 31(2), 813–825 (2012)
15. Kaminsky, E., Grois, D., Hadar, O.: Dynamic Computational Complexity and Bit Allocation for Optimizing H.264/AVC Video Compression. Journal of Visual Communication and Image Representation 19(1), 56–74 (2008)
16. Saha, A., Mukherjee, J., Sural, S.: A Neighborhood Elimination Approach for Block Matching in Motion Estimation. Signal Processing, Image Communication 26(8), 438–454 (2011)

NeuCube EvoSpike Architecture for Spatio-temporal Modelling and Pattern Recognition of Brain Signals

Nikola Kasabov

Knowledge Engineering and Discovery Research Institute - KEDRI,
Auckland University of Technology, and Institute
for Neuroinformatics - INI, ETH and University of Zurich
nkasabov@aut.ac.nz
www.kedri.info,
ncs.ethz.ch/projects/evospike

Abstract. The brain functions as a spatio-temporal information processing machine and deals extremely well with spatio-temporal data. Spatio- and spectro-temporal data (SSTD) are the most common data collected to measure brain signals and brain activities, along with the recently obtained gene and protein data. Yet, there are no computational models to integrate all these different types of data into a single model to help understand brain processes and for a better brain signal pattern recognition. The EU FP7 Marie Curie IIF EvoSpike project develops methods and tools for spatio and spectro temporal pattern recognition. This paper proposes a new evolving spiking model called NeuCube as part of the EvoSpike project, especially for modeling brain data. The NeuCube is 3D evolving Neurogenetic Brain Cube of spiking neurons that is an approximate map of structural and functional areas of interest of an animal or human brain. Optionally, gene information is included in the NeuCube in the form of gene regulatory networks that relate to spiking neuronal parameters of interest. Different types of brain SSTD can be used to train a NeuCube, including: EEG, fMRI, video-, image- and sound data, complex multimodal data. Potential applications are: EEG -, fMRI-, and multimodal brain data modeling and pattern recognition; Brain-Computer Interfaces; cognitive and emotional robots; neuro-prosthetics and neuro-rehabilitation; modeling brain diseases. Analysis of the internal structure of the model can trigger new hypotheses about spatio-temporal pathways in the brain.

Keywords: evolving neurogenetic brain cube, spatio/spectro-temporal brain data, pattern recognition, spiking neural networks, gene regulatory networks, computational neuro-genetic modelling, probabilistic modeling, personalized modeling, EEG, fMRI.

1   Spatio/ Spectro-temporal Information Processes in the Brain

1.1   Spatio-temporal Information Processes in the Brain

The brain is a complex integrated spatio-temporal information processing machine. An animal or human brain has different functional areas that are spatially distributed

N. Mana, F. Schwenker, and E. Trentin (Eds.): ANNPR 2012, LNAI 7477, pp. 225–243, 2012.

in a constrained 3D space, e.g. fig.1. When the brain processes information, either triggered by external stimuli, or by inner processes, such as visual-, auditory-, somatosensory-, olfactory-, control-, emotional-, or all together, complex spatio-temporal paths are activated and patterns are formed across the whole brain, e.g. fig.2.

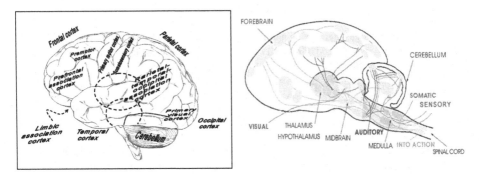

Fig. 1. Structural/functional areas of the brain are spatially distributed   **Fig. 2.** Spatio-temporal signaling paths in the brain

Example 1 (from [3]): '…the language task involves transfer of information from the inner ear through the auditory nucleus in thalamus to the primary auditory cortex (Brodmann's area 41), then to the higher-order auditory cortex (area 42), before it is relayed to the angular gyrus (area 39). Angular gyrus is a specific region of the parietal-temporal-occipital association cortex, which is thought to be concerned with the association of incoming auditory, visual and tactile information. From here, the information is projected to Wernicke's area (area 22) and then, by means of the *arcuate fasciculus*, to Broca's area (44, 45), where the perception of language is translated into the grammatical structure of a phrase and where the memory for word articulation is stored. This information about the sound pattern of the phrase is then relayed to the facial area of the motor cortex that controls articulation so that the word can be spoken. It turned out that a similar pathway is involved in naming an object that has been visually recognized. This time, the input proceeds from the retina and LGN (lateral geniculate nucleus) to the primary visual cortex, then to area 18, before it arrives to the angular gyrus, from where it is relayed by a particular component of arcuate fasciculus directly to Broca's area, bypassing Wernicke's area' [end of citation].

The example above shows that the brain processes information through the activation of complex spatio-temporal pathways involving many areas. Can this principle be modeled in a computer model, resulting in an improved accuracy of brain signal pattern recognition and new knowledge discovered?

Many other studies of spatio-temporal pathways in the brain have been conducted, e.g. birdsong learning [14].

1.2   Measuring Spatio/Spectro Temporal Brain Data

Much of the measured brain data is indeed spatio-temporal. The most common types are EEG [10,15,41] and fMRI data [58] (examples are shown in figs.3 and 4 respectively).

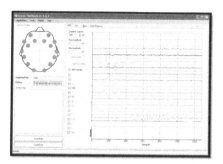
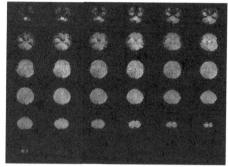

Fig. 3. EEG data recorded with the use of Emotive

Fig. 4. fMRI data (from http://www.fmrib.ox. ac.uk)

Many other techniques for collecting brain spatio-temporal data exist [19,9,6].

1.3   Spatio-temporal Brain Gene Data

Spatio-temporal activity in the brain depend on the internal brain structure, on the external stimuli and also – very much on the dynamics at the gene-protein level. The Allen Human Brain Atlas (www.brain-map.org) of the Allen Institute for Brain Science (www.alleninstitute.org) has shown that at least 82% of the human genes are expressed in the brain. For 1000 anatomical sites of human brains 100 mln data points are collected that indicate gene expressions of each of the genes and underly the biochemistry of the sites. Different genes express as mRNA, microRNA and proteins differently in different areas of the brain and are involved in all information processes, from simple spiking activity, to perception, decision making and emotions. For example, in [66] it is suggested that both the firing rate (rate coding) and spike timing depends on microRNAs that play a role in fast and slow dynamics and adaptive sensorimotor responses controlled by the cerebellar nuclei. Spatio-temporal patterns of population of Purkinji cells are shaped by activities in the molecular layer of interneurons. Functional connectivity develops in parallel with structural connectivity during brain maturation - a growth-elimination process (synapses are created and eliminated) depending on gene expression. Postsynaptic AMPA-type glutamate receptors (AMPARs) mediate most fast excitatory synaptic transmissions and are crucial for many aspects of brain function, including learning, memory and cognition [25,3]. The spiking activity of a neuron may affect as a feedback the expressions of genes [66]. As pointed in [66] on a longer time scale of minutes and hours the function of neurons may cause changes of the expression of hundreds of genes transcribed into mRNAs and also in microRNAs, which makes the short-term, the long-term and the genetic memories of a neuron linked together in a global memory of the neuron and further - of the whole neural system.

All the above facts and the numerous studies of gene activities related to brain spiking activity suggest that genes have to be taken into account when modeling brain data.

2   Spiking Neural Networks for SSTD Modelling and Pattern Recognition

The human brain has the amazing capacity to learn and recall patterns at different time scales, ranging from milliseconds, to years and possibly to millions of years (e.g. genetic information, accumulated through evolution). Thus the brain is the ultimate inspiration for the development of new machine learning techniques for brain SSTD modeling. Indeed, brain-inspired Spiking Neural Networks (SNN) [26,27,8,20,21] have the potential to learn SSTD by using trains of spikes (binary temporal events) transmitted among spatially located synapses and neurons. Both spatial and temporal information can be encoded in an SNN as locations of synapses and neurons and time of their spiking activity respectively.

Models of single neurons as well as computational SNN models, along with their respective applications, have been already developed [1,5,7,35,42,65], including evolving connectionist systems and evolving spiking neural networks (eSNN) in particular [35]. eSNN learn data incrementally by one-pass propagation of the data via creating and merging spiking neurons [35,65]. In [65] an eSNN is designed to capture features and to aggregate them into audio and visual perceptions for the purpose of person authentification.

Recently some new techniques have been developed as part of the EvoSpike project (http://ncs.ethz.ch/projects/evospike) and other projects that allow the creation of new types of computational models to deal with SSTD [40], e.g.: SPAN [45,46]; deSNN [13]; reservoir eSNN   [47,55,57]. Applications included moving object recognition [40,13,34], sign language recognition [53], EEG pattern recognition [48].

Still, the current methods and systems for SSTD do not reflect on the brain structure and functionality which may be beneficial when modeling brain SSTD, for brain pattern recognition and consequently for understanding brain spatio-temporal connectivity. There are some functional models of the brain, such as Izhikevich's [33] and the IBM Modha's models, but they do not suggest methods for brain SSTD pattern recognition, neither they include a broader genetic information.

The proposed here NeuCube Architecture is aiming at bringing the two research lines together through the creation of SNN that reflect on the brain structure and functionality with the hypothesis that it can be efficiently used for brain SSTD modeling and pattern recognition along with the discovery of new knowledge about brain spatio-temporal connectivity under different conditions.

3   The Proposed NeuCube Architecture

3.1   Overall Architecture and Main Principles

The main idea is to support the creation of multi-modular integrated systems, where different modules, consisting of different neuronal types and genetic parameters *correspond* in a way to different parts of the brain and different functions of interest (e.g.: vision; sensory information processing; sound recognition; motor-control) and the whole system works in an integrated mode for brain signal pattern recognition. A concrete model, built with the use of the NeuCube architecture, would have a specific structure and a set of algorithms depending on the problem and the application, e.g.: classification of EEG signals; recognition of fMRI data; BCI; emotional cognitive robotics; modeling AD.

A block diagram of the general EvoSpike NeuCube architecture is shown in fig.5. It consists of the following modules: input information encoding module; NeuCube module; output module; gene regulatory networks (GRN) module. The input module transfers input data into trains of spikes when sound/image or other sensory data is processed. EEG-, fMRI and other brain data is directly entered into the main module – the NeuCube.

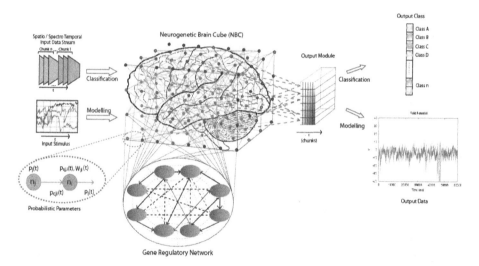

Fig. 5. A schematic diagram of a general NeuCube architecture, consisting of: input encoding module; NeuCube module; output function module; gene regulatory networks (GRN) module (optional)

The NeuCube is an approximate map of relevant for the study brain regions, along with relevant genetic information, as a 3D spiking neuronal structure. An initial NeuCube structure can include known connections between different areas of the brain. There are two types of SSTD used for both training a particular NeuCube and to recall it on new data:

(a) SSTD, measuring the activity of the brain (e.g. EEG, fMRI) when certain stimuli are presented. This data is entered to corresponding spatially located areas of the Cube;

(b) Direct stimuli data, e.g. sound, spoken language; video data; tactile data; odor data etc., which are first encoded into spike trains in the input module.

A NeuCube architecture is three-tire, consisting of a NeuCube module at the middle level, gene regulatory networks (GRN) at the lowest level, and a higher level evaluation (classification) module. The GRN control parameters of neurons from the NeuCube. Genes from the GRN are also affected by spiking activity of these neurons. Neurons from the NeuCube are connected to neurons of the output module in a two-way mode, so that the state of the Cube can be recognized/classified/interpreted in the Output Module and the result of this can also influence further activity of the neurons in the Cube as a feedback.

Different types of neurons and learning rules can be used in different areas of the architecture which is evolving. While a NeuCube Architecture has an initial structure, new neurons and connections are created from incoming data when this is needed. The structure and the functionality of a NeuCube Architecture evolve in time from incoming data.

Learning in the NeuCube Architecture is performed in two stages:

- Unsupervised training of a NeuCube, where SSTD is entered into relevant areas of the cube over time and unsupervised learning is used to establish the connection weights. The NeuCube will learn to activate similar spiking trajectories when similar input stimuli are presented – a *polychronisation* effect [32].

- Supervised training of the output module, where the same SSTD used for training is now propagated again through the trained NeuCube and output neurons are trained to classify the state of the cube into pre-defined labels (or output spike sequences). As a special case, all neurons from the NeuCube are connected to every output neuron. Feedback connections from output neurons to neurons in the NeuCube can be created for reinforcement learning.

Memory of a NeuCube Architecture is represented as a combination of: (a) Short-term memory, represented as changes of the PSP and temporary changes of synaptic efficacy; (b) Long-term memory, represented as a stable establishment of synaptic efficacy; (c) Genetic memory, represented as a change in the genetic code and the gene/ protein expression level as a result of the above short-term and long term memory changes and evolutionary processes.

More detailed description of a realization of the NeuCube Architecture is given below.

3.2   Input Encoding Module

This module converts input data into trains of spikes. The spike trains will be directed to the corresponding areas of the NeuCube. Temporal spike coding (rather than rate coding) is used following biological facts (e.g. [50,51]). Two methods are used to transfer continuous SSTD into spike trains.

3.2.1   Population Rank Coding

Continuous value input data is transferred into spikes so that the current value of each input variable (e.g. pixel, EEG channel, voxel) is entered into a population of neurons that emit spikes based on their receptive fields [4-6] – fig.6a. This is suitable when the input data is presented as a sequence of frames. If this is a frame (a vector) of EEG data of 64 channels, for example, each channel value is entered into a population of spiking neurons from the NeuCube module that are spatially located. These neurons transfer the channel value into a sequence of spikes generated based on the membership of this value to the receptive fields of the neurons (fig.6a). If there is no sufficient membership degree to any of the existing neurons, new neurons are generated (evolved) to accommodate this value.

3.2.2   Address Event Representation (AER)

This is based on thresholding the difference between two consecutive values of the same input variable over time as it is in the artificial retina [12] - fig.6b. This is suitable when the input data is a stream and only the changes in consecutive values are processed.

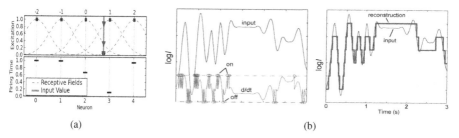

(a)   (b)

Fig. 6. (a) Population rank order coding of input information; (b) Address Event Representations (AER) of the input information (e.g. [12])

The input information is entered either on-line (for on-line, real time applications) or as a batch data. The *time* of the input data is in principal different from the internal *time* of information processing in the NeuCube.

3.3   The NeuCube Module

3.3.1   The Structure

The structure of the NeuCube is a 3D approximate map of brain areas of interest. Small world connections are used to initialize the connections, where neurons within a functional area of interest from the cube (e.g. visual area) are more densely connected than neurons across areas, still depending on the distance between the neurons [62]. Prior connections can be set based on prior knowledge. New neurons are created to accommodate data that is not possible to accommodate in the existing neurons following the ECOS principle [35]. The new neurons are connected with the rest following initially the small-world principle.

Input information in the NeuCube is entered not to arbitrarily selected neurons (as it is in the reservoir computing [42,62]), but to context dependent one from the NeuCube, e.g. visual information is entered into the neurons in the NeuCube that correspond to the visual cortex. In this way the cube has a meaningful structure that can be understood in terms of spatio-temporal pathways.

In the NeuCube each area can in principally use different neuronal models and learning algorithms. As a first implementation here, we use a probabilistic leaky integrate and fire model (LIFM) [39] and STDP learning rule [59].

3.3.2   The Neuronal Model

Fig. 6c and d illustrate the structure and the functionality of the LIFM [19-21]. The neuronal post synaptic potential (PSP), also called membrane potential u(t), increases with every input spike at a time t, multiplied to the synaptic efficacy (strength), until it reaches a threshold θ. After that, an output spike is emitted and the membrane potential is reset to an initial state. Between spikes, the membrane potential leaks, which is defined by a temporal parameter τ.

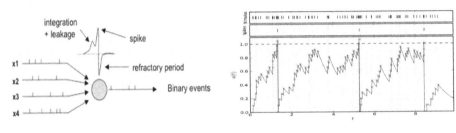

Fig. 6. (c) The structure of the LIFM   (d) functionality of the LIFM

In the probabilistic neuronal model [39] in addition to connection weights $w_{j,i}(t)$, a probabilistic spiking neuron model has the following three probabilistic parameters:

- A probability $p_{cj,i}(t)$ that a spike emitted by neuron n_j will reach neuron n_i at a time moment t through the connection between n_j and n_i. If $p_{cj,i}(t)=0$, no connection and no spike propagation exist between neurons n_j and n_i. If $p_{cj,i}(t) = 1$ the probability for propagation of spikes is 100%.

- A probability $p_{sj,i}(t)$ for the synapse $s_{j,i}$ to contribute to the PSPi(t) after it has received a spike from neuron n_j.

- A probability $p_i(t)$ for the neuron n_i to emit an output spike at time t once the total PSP_i (t) has reached a value above the PSP threshold (a noisy threshold).

The total PSPi(t) of the probabilistic spiking neuron n_i is now calculated using the following formula [39]:

$$PSP_i(t) = \sum_{p=t_0,..,t} \; (\sum_{j=1,..,m} e_j f_1(p_{cj,i}(t-p))f_2(p_{sj,i}(t-p))w_{j,i}(t)+\eta(t-t_0))$$

where: e_j is 1, if a spike has been emitted from neuron n_j, and 0 otherwise; $f_1(p_{cj,i}(t))$ is 1 with a probability $p_{cji}(t)$, and 0 otherwise; $f_2(p_{sj,i}(t))$ is 1 with a probability $p_{sj,i}(t)$, and 0 otherwise; t_0 is the time of the last spike emitted by n_i; $\eta(t-t_0)$ is an additional term representing decay in the PSP_i. As a special case, when all or some of the probability parameters are fixed to "1", the above probabilistic model will be simplified and will resemble the well known IFM.

The probabilistic model of a neuron is flexible to represent different types of neuronal models depending on the probabilistic parameters chosen, along with being suitable for the functional link with genes and GRN [39].

3.3.3   The Learning Rule

It has been demonstrated that a SNN that utilises probabilistic neuronal model can learn better SSTD than traditional SNN with simple IFM, especially in a nosy environment [53-56]. The effect of each of the above three probabilistic parameters on the ability of a SNN to process noisy and stochastic information was studied in [57,55].

The STDP learning rule [59], used here to train the NeuCube, utilizes Hebbian plasticity [24] in the form of long-term potentiation (LTP) and depression (LTD) (fig.7a). Efficacy of synapses is strengthened or weakened based on the timing of post-synaptic action potential in relation to the pre-synaptic spike. If the difference in the spike time between the pre-synaptic and post-synaptic neurons is negative (pre-synaptic neuron spikes first) then the connection weight between the two neurons increases, otherwise it decreases. Through STDP, connected neurons learn consecutive temporal associations from data and new connections are also evolved. Pre-synaptic activity that precedes post-synaptic firing can induce long-term potentiation (LTP), reversing this temporal order causes long-term depression (LTD). Other spike time unsupersvised learning rules can be in principle used instead of STDP, depending on the problem in hand.

3.3.4   Evolvability of the NeuCube

Initial structure of the NeuCube is preliminary defined based on the brain data available and the problem. But the structure is also evolving through the creation of new neurons and new connections based on the ECOS principles [35]. If new data do not activate sufficiently existing neurons, new neurons are created and allocated to match the data along their new connections.

3.4   The Output Module

During the training of the NeuCube with SSTD, a mapping of the input data into spatio-temporal patterns of connectivity and spiking activity pathways of the NeuCube is learned. After training, these NeuCube patterns are learned to be associated with known classes in a classifier of the Output Module, or actions are produced for motor control tasks. Feedback connections from the Output Module to the NeuCube are possible to establish for reinforcement learning.

In our first realization, all spiking neurons from the NeuCube are connected to each of the output neurons. Two different methods are developed in the EvoSpike project: deSNN for classification of NeuCube states; SPAN for generating motor control signals in response to certain patterns of activity of the NeuCube.

3.5.1   deSNN for Fast, On-Line Classification Based on AER Principle

The dynamic eSNN (deSNN) [13] combines rank-order and temporal (e.g. STDP or SDSP) learning rules. The initial values of synaptic weights are set according to the rank-order learning assuming the first spikes in the NuCube are more important than the rest. The weights are further modified to accommodate following spikes from the NeuCube activated by the same SSTD stimulus pathway, with the use of a temporal learning rule – STDP (or SDSP).

The rank-order learning rule [61,60] uses important information from the input spike trains – the rank of the first incoming spikes on the neuronal synapses (fig.7b). It establishes a priority of inputs (synapses) based on the order of the spike arrival on these synapses for a particular pattern, which is a phenomenon observed in biological systems as well as an important information processing concept for some problems, such as computer vision and control [4,61,51]. This type of learning has several advantages when used in SNN, mainly: fast, one- pass, learning (as it uses the extra information of the order of the incoming spikes) and asynchronous data entry (synaptic inputs are accumulated into the neuronal membrane potential in an asynchronous way). The postsynaptic potential of a neuron i at a time t is calculated as:

$$PSP(i,t) = \sum mod^{order(j)} w_{j,i}$$

where: *mod* is a modulation factor; j is the index for the incoming spike at synapse j,i and $w_{j,i}$ is the corresponding synaptic weight; *order(j)* represents the order (the rank) of the spike at the synapse j,i among all spikes arriving from all m synapses to the neuron i. The *order(j)* has a value 0 for the first spike and increases according to the input spike order. An output spike is generated by neuron i if the PSP (i,t) becomes higher than a threshold PSP_{Th} (i). During the training process of a classifier, for each training input pattern (e.g. a spiking trajectory of the NeuCube) the connection weights are calculated based on the order of the incoming spikes [61,65]: $\Delta w_{j,i}$ (t)= mod $^{order (j,i (t))}$

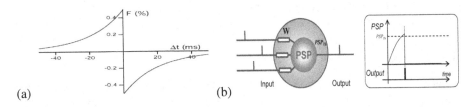

(a)   (b)

Fig. 7. a,b: (a) An example of synaptic change in a STDP learning neuron [59]; (b) Rank-order learning neuron

3.4.2   The SPAN Algorithm for Spike Pattern Association and Motor Control Applications

SPAN is an algorithm for both classification and spike pattern association [44-46]. The whole spiking pattern of the NeuCube is learned here, rather than spike by spike as it is in the deSNN. Delta rule is used for the purpose [64]. SPAN learns to generate an output spike at a certain time, or a pattern of temporally distribute spikes over time, when a certain input spatio-temporal pattern of spikes (e.g. spiking sequence from the NeuCube) is recognized. Two algorithms are developed, one for batch mode learning [46], and one for incremental learning [45]. SPAN is a suitable algorithm for *control applications*, so that when a certain pattern is recognized, a series of motor control signals are emitted at different times on one or on several outputs. In principle, SPAN is a further development of a class of algorithms, including: ReSuMe [49]; Chronotron [16]; Tempotron [23].

4   The Gene Regulatory Network (GRN) Module

4.1   How Do Genes Affect the Spiking Activity of the NeuCube Architecture?

The GRN module from fig.5 uses as a main principle the analogy with biological facts about the relationship between spiking activity and gene/protein dynamics in order to control the learning and spiking parameters in a SNN. Biological support of this can be found in numerous publications (e.g. [3,25,66]). We illustrate this with the following example.

Example 2: In [38] the dramatic effect of a change of single gene, that regulates the τ parameter of the LIF neurons, on the spiking activity of the whole SNN cube of 1000 neurons, is shown in fig.8.

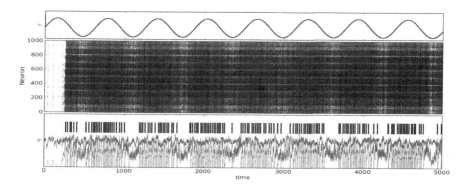

Fig. 8. A single gene expression level over time can affect the pattern of activity of a whole cube of 1000 neurons (from [38]). The gene controls the τ parameter of all 1000 LIF neurons over a period of five seconds. The top diagram shows the evolution of τ. The response of the 10000 spiking neurons is shown as a raster plot of spike activity. A black point in this diagram indicates a spike of a specific neuron at a specific time in the simulation. The bottom diagram presents the evolution of the membrane potential of a single neuron from the network (green curve) along with its firing threshold ϑ (red curve). Output spikes of the neuron are indicated as black vertical lines in the same diagram (from [38]).

4.2   A Neuro-genetic Model of a Neuron

The GRN module has several GRNs, each controlling parameters of respective functional area of the NeuCube. The question how to relate the genes from a GRN to the parameters of the spiking neurons that control their spiking activity is to be addressed.

In our realisation here we use the neurogenetic model as an extension of the LIF and the probabilistic neuronal model. Here, instead of using two types of synapses – inhibitory and excitatory as it is in almost all SNN models, we use the proposed in [36,3] four types of synapses for *fast excitation, fast inhibition, slow excitation, and slow inhibition*. The contribution of each to the PSP of the neurons in the NeuCube is defined by the level of expression of different genes/proteins along with the presented external stimuli. The model utilises known information about how proteins and genes affect the spiking activities of a neuron. Table 1 shows what proteins affect each of the four types of synapses in our neurogenetic model. For a real case application, apart from the GABAB receptor some other metabotropic and other receptors could be also included. This information is used to calculate the contribution of each of the *four* different synapses, connected to a neuron n_i, to its post synaptic potential PSPi(t):

$$ \varepsilon_{ij}^{synapse}(s) = A^{synapse} \left(\exp\left(-\frac{s}{\tau_{decay}^{synapse}}\right) - \exp\left(-\frac{s}{\tau_{rise}^{synapse}}\right) \right) $$

where: $\tau_{decay/rise}^{synapse}$ are time constants representing the rise and fall of an individual synaptic PSP; A is the PSP's amplitude; $\varepsilon_{ij}^{synapse}$ represents the type of activity of the synapse between neuron j and neuron i that can be measured and modelled separately for a fast excitation, fast inhibition, slow excitation, and slow inhibition (it is affected by different genes/proteins). External inputs can also be added to model background noise, background oscillations or environmental information.

The genes that relate to the parameters of the neurons are also related to the activity of other genes, forming a GRN. A GRN defines the dynamic interactions between genes/proteins over time that affect the spiking activity of the neuron. Although biologically plausible, a GRN model is only a highly simplified general model that does not necessarily take into account the exact chemical and molecular interactions. A GRN model is defined by:

(a)   a set of genes/proteins, $G= (g_1, g_2, \ldots, g_k)$;

(b)   an initial state of the level of expression of the genes/proteins $G(t=0)$;

(c)   an initial state of a connection matrix $L = (L_{11}, \ldots, L_{kk})$, where each element L_{ij} defines the known level of interaction (if any) between genes/proteins g_j and g_i;

(d)   activation functions f_i for each gene/protein g_i from G. This function defines the gene/protein expression value at time (t+1) depending on the current values $G(t)$, $L(t)$ and some external information $E(t)$: $g_i(t+1)= f_i (G(t), L(t), E(t))$.

Table 1. Neuronal action potential parameters and related proteins and ion channels in the computational neuro-genetic model of a spiking neuron: AMPAR - (amino- methylisoxazole-propionic acid) AMPA receptor; NMDR - (N-methyl-D-aspartate acid) NMDA receptor; $GABA_AR$ - (gamma-aminobutyric acid) $GABA_A$ receptor, $GABA_BR$ - $GABA_B$ receptor; SCN - sodium voltage-gated channel, KCN - kalium (potassium) voltage-gated channel; CLC - chloride channel (from [3, 36]))

Different types of action potential of a spiking neuron	Related neurotransmitters and ion channels
Fast excitation PSP	AMPAR [25]
Slow excitation PSP	NMDAR
Fast inhibition PSP	$GABA_AR$
Slow inhibition PSP	$GABA_BR$
Modulation of PSP	mGluR
Firing threshold	Ion channels SCN, KCN, CLC

The neurogenetic model of a neuron is a further extension of the probabilistic model from section 3 that is our generic model for the NueCube architecture with the following features: LIF type activity; four different synapses regulated by genes from a GRN; three probabilistic parameters attached to the connections, to the synaptic activities and the activation threshold respectively.

4.3   Optimisation of the GRN Model and the NeuCube Parameters

A major problem with the NeuCube architecture from fig.5 is how to optimize the numerous parameters of the model. Several evolutionary computation methods have been developed for this purpose by our team: PSO [44]; quantum inspired genetic algorithm [11,54,56]; quantum inspired PSO [47,53]. The quantum inspired evolutionary methods can deal with a very large dimensional space as each quantum-bit chromosome represents the whole space, each point to certain probability [35,39]. Such algorithms are faster and lead to a close solution to the global optimum in a very short time. We will be using same parameter values (same GRN) for all neurons in a functional area of the NeuCube. That will result in a significant reduction of the parameters to be optimized. This can be interpreted as 'average' parameter value for the neurons in the same area.

When defining parameters of the NeuCube, prior knowledge about the association of spiking parameters with relevant genes/proteins (neuro-transmitter, neuro-receptor, ion channel, neuro-modulator) as described in [37] can be also used.

5   Brain SSTD Modelling and Pattern Recognition

5.1   EEG-, fMRI-, and Integrated EEG + fMRI Pattern Recognition

Recorded brain signals (EEG, fMRI, combined, other) can be used to build a NeuCube model in the following way:

5.1.1   Training of a NeuCube Architecture

1. The Cube is spatially structured to match the spatial distribution of the EEG-, or fMRI- data, or both.
2. The available data is entered into the NeuCube, so that data is entered into the corresponding neuron (neurons) in the Cube. STDP learning is applied in the NeuCube to establish the connection weights of spatial-temporal patterns of pathway connectivity.
3. The output classification module (control module) is trained to recognize the states of the NeuCube into predefined classes (activate desired control devices).
4. The polychronisation patterns formed in the NeuCube, that correspond to each input data pattern, are analysed for the purpose of understanding the data.

5.1.2   Recall of the Trained NeuCube Architecture on New Data

1. New input data is propagated through the NeuCube after encoding into spike trains, so that data from particular areas of the brain are submitted to corresponding neurons in the NeuCube.
2. Output classification (control) results are recorded.
3. The activity of the NeuCube in terms of polychronisation trajectories are analysed and conclusions are made regarding the new input data and the spatio-temporal connectivity and pathways.

5.1.3   Further Adaptation of the NeuCube Architecture

If new data is available that belongs to some of the existing or new classes, further training of the NeuCube architecture is applied and new output classification (control) neurons are added/evolved and trained in the same way as in procedure 5.1.1.

5.1.4   Using EEG, fMRI and Multimodal Data in one NeuCube Architecture.

The proposed here NeuCube architecture allows for multimodal brain data to be used for training the system. In this respect, EEG and fMRI data along with multimodal (e.g. sound/vision/olfactory) data can be used one after another to train the NeuCube, with the hypothesis that each data set will enhance the learning and recognition of the other, making use of the specific characteristics of each of the data sets, such as precise spatial information (fMRI) and precise temporal information (EEG) are used together.

5.1.5   Applications

EEG pattern recognition [48] can be directed to practical applications, such as: BCI [31]; classification of epilepsy [22] (e.g., using deSNN); robot control through EEG signals [41] and robot navigation [2] (e.g., using SPAN). In case of BCI, one scenario is when a NeuCube architecture can be evolved and dynamically visualized to represent the learning process of a subject. fMRI and combined data can be used to discover spatio-temporal connectivity pathways in the brain under different conditions.

5.2   Neuro-genetic Modelling for Cognitive/Emotional Systems and Brain Diseases

Building artificial cognitive systems, such as robots, that are able to communicate with humans in a human-like way has been a goal for computer scientists for decades now. Cognition is closely related with emotions. Basic emotions are joy, sadness, anger, fear, disgust, surprise, but other human emotions play role in cognition as well (pride, shame, regret, etc.). Gene parameters and gene regulatory network models can be used to help a robot evolve its structure and functionality, where genes represented either parameters of dynamic learning processes of the robot, or parameters that connected dynamically different functional modules of the robot (e.g. see [2,43]).

The proposed here NeuCube architecture would make it possible to model cognition-emotion processes that would further enable us to create more human-like robotic systems. Creating human-like behavior in a cognitive system requires understanding relevant brain processes at different levels of information processing. For example, it is known that human emotions depend on the expression and the dynamic interaction of neuromodulators (serotonin, dopamine, noradrenalin and acetylcholine) and some other relevant genes and proteins (e.g., 5-HTTLRP, DRD4, DAT), that are functionally linked to the spiking activity of the neurons in certain areas of the brain and participate in a GRN. They have wide ranging effects on brain functions and their relationship to emotional states is well established. Noradrenaline is important to arousal and attention mechanisms, acetylcholine has a key role in encoding memory function, dopamine is related to aspects of learning and reward seeking behavior and may signal probable appetitive outcome, whereas serotonin release may inhibit behavior with probable aversive outcome. Reward and punishment are fundamental features of emotional behavior.

The NeuCube architecture can be potentially used to model brain data to enhance the study of the brain in its complexity and different levels of operation related to learning, memory, emotions, cognition, or brain diseases. Based on prior information and available data, different NeuCube architectures can be developed for the study of various brain states, conditions and diseases, including [3,18]: epilepsy [22]; schizophrenia; mental retardation; brain aging and AD [52,37], Parkinson disease; clinical depression.

6   Conclusion and Further Directions

The presented here NeuCube architecture opens opportunities for the development of new generation of brain-like intelligent systems for pattern recognition and for a better understanding of the spatial-temporal connectivity of the brain. Software simulations of different brain data will be conducted in the future, including: EEG pattern recognition and comparing results with previous studies of using reservoir computing [57]; EEG pattern recognition for BCI [41,48]; fMRI pattern recognition and comparison with previous studies [58]. A module for dynamic visualisation of NeuCube polychronisation patterns will be developed along with their interpretation in terms of functional and structural pathways discovered. Potential practical applications would include: neuro-rehabilitation robots [63]; neuro-prosthetics;

control and navigation of wheelchair [10]; cognitive and emotional systems [37]; serious games. Implementing the NeuCube architecture, based on the neurogenetic model of neuron with four types of synapses and probabilistic parameters, as a neuromorphic chip [28-30,17] is a next step in this direction. It will make a large scale of engineering applications of this model possible.

Acknowledgement. This project is supported by the EU FP7 Marie Curie project EvoSpike PIIF-GA-2010-272006, hosted by the Institute for Neuroinformatics at ETH/UZH Zurich (http://ncs.ethz.ch/projects/evospike), and also by the Knowledge Engineering and Discovery Research Institute (KEDRI, http://www.kedri.info) of the Auckland University of Technology and the New Zealand Ministry of Science and Innovation. I have discussed issues related to this project with: A.Mohemmed, S.Schliebs, G.Indivery, K.Dhoble (who also drew fig.5), N.Nuntalid, H.Nuzlu, N.Murli, R.Alkatel, N.Scott. And thanks to editors Mana, Scwenker and Trentin for giving me the opportunity to publish this model at its preliminary stage of development.

References

1. Belatreche, A., Maguire, L.P., McGinnity, M.: Advances in Design and Application of Spiking Neural Networks. Soft Comput. 11(3), 239–248 (2006)
2. Bellas, F., Duro, R.J., Faiña, A., Souto, D.: MDB: Artificial Evolution in a Cognitive Architecture for Real Robots. IEEE Transactions on Autonomous Mental Development 2, 340–354 (2010)
3. Benuskova, L., Kasabov, N.: Computational neuro-genetic modelling, 290 pages. Springer, New York (2007)
4. Berry, M.J., Warland, D.K., Meister, M.: The structure and precision of retinal spiketrains. PNAS 94(10), 5411–5416 (1997)
5. Bohte, S., Kok, J., LaPoutre, J.: Applications of spiking neural networks. Information Processing Letters 95(6), 519–520 (2005)
6. Bohte, S.M.: The evidence for neural information processing with precise spike-times: A survey. Natural Computing 3 (2004)
7. Brette, R., et al.: Simulation of networks of spiking neurons: a review of tools and strategies. J. Comput. Neuroscience 23, 349–398 (2007)
8. Buonomano, D., Maass, W.: State-dependent computations: Spatio-temporal processing in cortical networks. Nature Reviews, Neuroscience 10, 113–125 (2009)
9. De Zeeuw, C.I., Hoebeek, F.E., Bosman, L.W.J., Schonewille, M.: Spatiotemporal firing patterns in the cerebellum. Nature Reviews Neuroscience 12, 327–344 (2011), doi:10.1038/nrn3011
10. Craig, D.A., Nguyen, H.T.: Adaptive EEG Thought Pattern Classifier for Advanced Wheelchair Control. In: Engineering in Medicine and Biology Society- EMBS 2007, pp. 2544–2547 (2007)
11. Defoin-Platel, M., Schliebs, S., Kasabov, N.: Quantum-inspired Evolutionary Algorithm: A multi-model EDA. IEEE Trans. Evolutionary Computation 13(6), 1218–1232 (2009)
12. Delbruck, T.: jAER open source project (2007),
   http://jaer.wiki.sourceforge.net

13. Dhoble, K., Nuntalid, N., Indivery, G., Kasabov, N.: Online Spatio-Temporal Pattern Recognition with Evolving Spiking Neural Networks utilising Address Event Representation, Rank Order, and Temporal Spike Learning. In: Proc. IJCNN 2012, Brisbane, pp. 554–560. IEEE Press (June 2012)
14. Doya, K., Sejnowski, T.: A Computational Model of Avian Song Learning. In: Gazzaniga, M. (ed.) The New Cognitive Neuroscience, pp. 469–482. MIT Press
15. Ferreira, A., Almeida, C., Georgieva, P., Tomé, A., Silva, F.: Advances in EEG-based Biometry. In: Campilho, A., Kamel, M. (eds.) ICIAR 2010. LNCS, vol. 6112, pp. 287–295. Springer, Heidelberg (2010)
16. Florian, R.V.: The chronotron: a neuron that learns to fire temporally-precise spike patterns
17. Furber, S., Temple, S.: Neural systems engineering, Interface. J. of the Royal Society 4, 193–206 (2007)
18. Gene and Disease, NCBI (2005), http://www.ncbi.nlm.nih.gov
19. Gerstner, W.: Time structure of the activity of neural network models. Phys. Rev. 51, 738–758 (1995)
20. Gerstner, W.: What's different with spiking neurons? In: Mastebroek, H., Vos, H. (eds.) Plausible Neural Networks for Biological Modelling, pp. 23–48. Kluwer Academic Publishers (2001)
21. Gerstner, W., Kreiter, A.K., Markram, H., Herz, A.V.M.: Neural codes: firing rates and beyond. Proc. Natl. Acad. Sci. USA 94(24), 12740–12741 (1997)
22. Ghosh-Dastidar, S., Adeli, H.: Improved Spiking Neural Networks for EEG Classification and Epilepsy and Seizure Detection. Integrated Computer-Aided Engineering 14(3), 187–212 (2007)
23. Gutig, R., Sompolinsky, H.: The tempotron: a neuron that learns spike timing-based decisions. Nat. Neurosci. 9(3), 420–428 (2006)
24. Hebb, D.: The Organization of Behavior. John Wiley and Sons, New York (1949)
25. Henley, J.M., Barker, E.A., Glebov, O.O.: Routes, destinations and delays: recent advances in AMPA receptor trafficking. Trends in Neuroscience 34(5), 258–268 (2011)
26. Hodgkin, A.L., Huxley, A.F.: A quantitative description of membrane current and its application to conduction and excitation in nerve. Journal of Physiology 117, 500–544 (1952)
27. Hopfield, J.J.: Neural networks and physical systems with emergent collective computational abilities. PNAS USA 79, 2554–2558 (1982)
28. Indiveri, G., Linares-Barranco, B., Hamilton, T., Van Schaik, A., Etienne-Cummings, R., Delbruck, T., Liu, S., Dudek, P., Häfliger, P., Renaud, S., et al.: Neuromorphic silicon neuron circuits. Frontiers in Neuroscience 5 (2011)
29. Indiveri, G., Chicca, E., Douglas, R.J.: Artificial cognitive systems: From VLSI networks of spiking neurons to neuromorphic cognition. Cognitive Computation 1(2), 119–127 (2009)
30. Indiveri, G., Stefanini, F., Chicca, E.: Spike-based learning with a generalized integrate and fire silicon neuron. In: 2010 IEEE Int. Symp. Circuits and Syst. (ISCAS 2010), Paris, May 30- June 02, pp. 1951–1954 (2010)
31. Isa, T., Fetz, E.E., Muller, K.: Recent advances in brain-machine interfaces. Neural Networks 22(9), 1201–1202 (2009)
32. Izhikevich, E.M.: Polychronization: Computation with Spikes. Neural Computation 18, 245–282 (2006)
33. Izhikevich, E.M., Edelman, G.M.: Large-Scale Model of Mammalian Thalamocortical Systems. PNAS 105, 3593–3598 (2008)

34. Kasabov, N., Dhoble, K., Nuntalid, N., Mohemmed, A.: Evolving Probabilistic Spiking Neural Networks for Spatio-temporal Pattern Recognition: A Preliminary Study on Moving Object Recognition. In: Lu, B.-L., Zhang, L., Kwok, J. (eds.) ICONIP 2011, Part III. LNCS, vol. 7064, pp. 230–239. Springer, Heidelberg (2011)

35. Kasabov, N.: Evolving connectionist systems: The knowledge engineering approach. Springer (2007); 1st edn. (2002)

36. Kasabov, N., Benuskova, L., Wysoski, S.: A Computational Neurogenetic Model of a Spiking Neuron. In: IJCNN 2005 Conf. Proc., vol. 1, pp. 446–451. IEEE Press (2005)

37. Kasabov, N., Schliebs, R., Kojima, H.: Probabilistic Computational Neurogenetic Framework: From Modelling Cognitive Systems to Alzheimer's Disease. IEEE Trans. Autonomous Mental Development 3(4), 1–12 (2011)

38. Kasabov, N., Schliebs, S., Mohemmed, A.: Modelling the Effect of Genes on the Dynamics of Probabilistic Spiking Neural Networks for Computational Neurogenetic Modelling. In: Proc. 6th Meeting on Computational Intelligence for Bioinformatics and Biostatistics, CIBB 2011, Gargangio, Italy, June 30-July 2. LNCS (LNBI). Springer (to appear, 2012)

39. Kasabov, N.: To spike or not to spike: A probabilistic spiking neuron model. Neural Netw. 23(1), 16–19 (2010)

40. Kasabov, N.: Evolving Spiking Neural Networks and Neurogenetic Systems for Spatio- and Spectro-Temporal Data Modelling and Pattern Recognition. In: Liu, J., Alippi, C., Bouchon-Meunier, B., Greenwood, G.W., Abbass, H.A. (eds.) WCCI 2012. LNCS, vol. 7311, pp. 234–260. Springer, Heidelberg (2012)

41. Lotte, F., Congedo, M., Lécuyer, A., Lamarche, F., Arnaldi, B.: A review of classification algorithms for EEG-based brain–computer interfaces. J. Neural Eng. 4(2), R1–R15 (2007)

42. Maass, W., Natschlaeger, T., Markram, H.: Real-time computing without stable states: A new framework for neural computation based on perturbations. Neural Computation 14(11), 2531–2560 (2002)

43. Meng, Y., Jin, Y., Yin, J., Conforth, M.: Human activity detection using spiking neural networks regulated by a gene regulatory network. In: Proc. Int. Joint Conf. on Neural Networks (IJCNN), Barcelona, pp. 2232–2237. IEEE Press (July 2010)

44. Mohemmed, A., Matsuda, S., Schliebs, S., Dhoble, K., Kasabov, N.: Optimization of Spiking Neural Networks with Dynamic Synapses for Spike Sequence Generation using PSO. In: Proc. Int. Joint Conf. Neural Networks, California, USA, pp. 2969–2974. IEEE Press (2011)

45. Mohemmed, A., Schliebs, S., Matsuda, S., Kasabov, N.: Evolving Spike Pattern Association Neurons and Neural Networks. Neurocomputing (in print)

46. Mohemmed, A., Schliebs, S., Matsuda, S., Kasabov, N.: SPAN: Spike Pattern Association Neuron for Learning Spatio-Temporal Sequences. International Journal of Neural Systems 22(4), 1–16 (2012)

47. Nuzly, H., Kasabov, N., Shamsuddin, S.: Probabilistic Evolving Spiking Neural Network Optimization Using Dynamic Quantum Inspired Particle Swarm Optimization. Australian Journal of Intelligent Information Processing Systems 11(1) (2010)

48. Nuntalid, N., Dhoble, K., Kasabov, N.: EEG Classification with BSA Spike Encoding Algorithm and Evolving Probabilistic Spiking Neural Network. In: Lu, B.-L., Zhang, L., Kwok, J. (eds.) ICONIP 2011, Part I. LNCS, vol. 7062, pp. 451–460. Springer, Heidelberg (2011)

49. Ponulak, F., Kasinski, A.: Supervised learning in spiking neural networks with ReSuMe: sequence learning, classification, and spike shifting. Neural Computation 22(2), 467–510 (2010) PMID:19842989

50. Reinagel, R., Reid, R.C.: Temporal coding of visual information in the thalamus. Journal of Neuroscience 20(14), 5392–5400 (2000)
51. Rokem, A., Watzl, S., Gollisch, T., Stemmler, M., Herz, A.V., Samengo, I.: Spike-timing precision underlies the coding efficiency of auditory receptor neurons. J. Neurophysiol. (2005)
52. Schliebs, R.: Basal forebrain cholinergic dysfunction in Alzheimer's disease – interrelationship with β-amyloid, inflammation and neurotrophin signaling. Neurochemical Research 30, 895–908 (2005)
53. Schliebs, S., Hamed, H.N.A., Kasabov, N.: Reservoir-Based Evolving Spiking Neural Network for Spatio-temporal Pattern Recognition. In: Lu, B.-L., Zhang, L., Kwok, J. (eds.) ICONIP 2011, Part II. LNCS, vol. 7063, pp. 160–168. Springer, Heidelberg (2011)
54. Schliebs, S., Kasabov, N., Defoin-Platel, M.: On the Probabilistic Optimization of Spiking Neural Networks. International Journal of Neural Systems 20(6), 481–500 (2010)
55. Schliebs, S., Mohemmed, A., Kasabov, N.: Are Probabilistic Spiking Neural Networks Suitable for Reservoir Computing? In: Int. Joint Conf. Neural Networks, IJCNN, San Jose, pp. 3156–3163. IEEE Press (2011)
56. Schliebs, S., Defoin-Platel, M., Worner, S., Kasabov, N.: Integrated Feature and Parameter Optimization for Evolving Spiking Neural Netw.: Exploring Heterogeneous Probabilistic Models. Neural Netw. 22, 623–632 (2009)
57. Schliebs, S., Nuntalid, N., Kasabov, N.: Towards Spatio-Temporal Pattern Recognition Using Evolving Spiking Neural Networks. In: Wong, K.W., Mendis, B.S.U., Bouzerdoum, A. (eds.) ICONIP 2010, Part I. LNCS, vol. 6443, pp. 163–170. Springer, Heidelberg (2010)
58. Sona, D., Veeramachaneni, S., Olivetti, E., Avesani, P.: Inferring Cognition from fMRI Brain Images. In: de Sá, J.M., Alexandre, L.A., Duch, W., Mandic, D.P. (eds.) ICANN 2007. LNCS, vol. 4669, pp. 869–878. Springer, Heidelberg (2007)
59. Song, S., Miller, K., Abbott, L., et al.: Competitive hebbian learning through spike-timing-dependent synaptic plasticity. Nature Neuroscience 3, 919–926 (2000)
60. Theunissen, F., Miller, J.P.: Temporal encoding in nervous systems: a rigorous definition. Journal of Computational Neuroscience 2(2), 149–162 (1995)
61. Thorpe, S., Gautrais, J.: Rank order coding. Computational Neuroscience: Trends in Research 13, 113–119 (1998)
62. Verstraeten, D., Schrauwen, B., D'Haene, M., Stroobandt, D.: An experimental unification of reservoir computing methods. Neural Networks 20(3), 391–403 (2007)
63. Wang, X., Hou, Z.G., Zou, A., Tan, M., Cheng, L.: A behavior controller for mobile robot based on spiking neural networks. Neurocomputing 71(4-6), 655–666 (2008)
64. Widrow, B., Lehr, M.: 30 years of adaptive neural networks: perceptron, madaline, and backpropagation. Proceedings of the IEEE 78(9), 1415–1442 (1990)
65. Wysoski, S., Benuskova, L., Kasabov, N.: Evolving spiking neural networks for audiovisual information processing. Neural Networks 23(7), 819–835 (2010)
66. Zhdanov, V.P.: Kinetic models of gene expression including non-coding RNAs. Phys. Rep. 500, 1–42 (2011)

Author Index